PensareDiverso Publishing
“The Field of Consciousness” Series.

Quantum Entanglement and the Field of Consciousness in Living Forms.

Bruno Del Medico.

The Biology of Fields

Quantum Entanglement and the Field of Consciousness in the New Biology of Life.

Implications for Health and Disease

Bruno Del Medico, Publisher.
PensareDiverso Editions
edizioni@delmedico.it
www.giardinoquantico.it

Table of Contents .

Quantum Entanglement and the Field of Consciousness in Living Forms. 2

Table of Contents 5

Introduction 13

Introduction. A New Conceptual Grammar. 15

Why life can no longer be conceived of as matter in motion. 16

The words with which we conceive of life. 18
From physical fields to fields of life. 20

Biology that works but doesn't understand. 23

Seeing is never neutral: images determine what life is. 25
Laboratories Without a Horizon: The Fragmentation of Biological Knowledge. 26
Predicting is not the same as understanding: the paradox of contemporary biology. 28

Rebellious Data: What Organisms Do That Theories Fail to Explain. 30

Emergence without mystery. 32
Between Expectations and Thresholds: Biology and Time That Doesn't Flow Linearly. 33
When the experiment succeeds, but the theory remains silent. 35
When we control the outcome but lose the meaning. 37
Restless Numbers: What Biological Data Do Not Reveal on Their Own. 39

Beyond the Gearwork: Why Organisms Are Not Well-Assembled Machines. 42

The hidden language of organisms...44
Information Without a Sender...47
Technology and Misunderstandings of the Living...........................49
From Calculation to Meaning: The New Unease in the Exact Sciences. 51

PART ONE. The Blind Spot of Modern Biology. 55

I. The Limits of Biochemical Explanation. 58

Circular causality in living systems: life that causes itself...............60
Words that mean more than they say. 62
Identity in Pieces: Why Life Cannot Be Taken Apart......................64
The Local Success of Biochemistry..................................67

From molecules to networks: when life changed scale.....................69
The whole that cannot be broken down. 72

The face we recognize without knowing why.................................74
When Cells "Know" What to Do: The Challenge of Purpose Without a Designer. 77

The purpose without a purpose: re-reading Aristotle in the light of contemporary life. 79
Diaries, letters, silences: the purpose that recurs in scientists' confessions...82
When systems do more than we programmed them to.....................84

II. When Evolution Plays, but Doesn't Write the Score....88

The Form That Endures: The Mystery of Biological Robustness.....90
Vortices, snowflakes, and cells: deceptive similarities in form. 92
When Form Demands a Deeper Principle..94
The Myth of the Sovereign Gene: When Information Is Not Enough to Explain Form. 96

The Gene and the Machine..98

The Gene on the Front Page: When the Media Turn Biology into Destiny. 100
To say "organism" or to say "mechanism": the moral consequences of a metaphor. 102
Changing form or giving form? 105
The role of attractors in theoretical biology. 107
Function comes later: because organisms are not well-designed tools. 110
Same architecture, a thousand uses: when form does not dictate destiny. 112
Because nature appears beautiful to us before it appears useful. 114
Showcases of Form: When Museums Taught Biology to See. 116
Scientific photography and the rediscovery of living form. 118
Natural aesthetics as a form of literacy in life. 120
The Stability of the Living: Why Form Demands a Deeper Theory. 122
PART TWO. The Invisible Thread of Life. 125
III. Coherence: The True Name of Life. 128
Living Time: How Coherence Creates Organisms' Internal Clocks. 130
Imperfect Order: Why Life Needs Noise. 132
Who governs whom? The decisive difference between automation and living organization. 134
What does "coherence" really mean? 137
When Order Becomes Physics: A Cultural History of a Key Term. 140
To measure is to intervene: when observation disrupts coherence. 143
The Test of Time: When Stability Becomes the True Sign of Life. 145
Coherence, synchrony, correlation, organization: four words, four different worlds. 147
Because to live means holding together what tends to fall apart. 150
Paying the price of unity: the energy cost of biological coherence. 152
No life is an island: coherence arises from dialogue with the environment. 154

When Unity Breaks Down: Death as the Decoherence of the Living. 156

More than the sum, less than the myth: coherence as a scientific fact. 159

Coherence as seen through mathematical models. 161
When Mathematics and Biology Learn to Understand Each Other. 163
The Shadow of Vitalism: How Emergence Escaped Scientific Suspicion. 165
This is what our reductionist training prevents us from recognizing. 167

IV. The Organism as a Non-Local System 170

Locality and Non-Locality: The Differences. 172
From the body to society: the non-locality that governs us 174
When the future determines the present. 175
When even machines become non-local. 176
When the whole was a nuisance to the parts. 177

Biological non-locality 179

Vital Tensions: The Silent Language of Tissues. 180
Green Thinking: How Plants Coordinate Without a Brain 182
Green Blindness: Why the West Has Been Slow to Understand Plants. 184
Do Plants "Know"? A New Ecology of Relationships 185
The Myth of Command: What Human Organizations Fail to Understand About Life 187

When everything influences everything: the end of the sovereign molecule. 190

We are not born in pieces: because in biology, the whole truly comes first. 193

PART THREE. Where Life Takes Center Stage. 196

V. What Is a Biological Field? 199

Biological field and environment: a necessary distinction. 201
Inhabiting the world, not merely enduring it: from the concept of "Umwelt" to the biological field. 203

The form that sustains itself: development, stability, and fields of life. 206
Attractive forms: why life chooses some paths over others. 208
From order to meaning: how the field transforms information into living form. 210
From the magnet to the organism: a brief history of an idea called the field. 212
Does a field really exist? Scientific realism put to the test by the invisible. 214
Because "field" never goes out of style: a word that captures complexity. 216
Why a "field," specifically? It is the right word to describe organized life. 218
What a biological field is not. 220
From Equilibrium to Sign: Slow Histories of Difficult Ideas. 222
The field as a structure of possibility. 225
VI. Information Taking Shape. 228
Growing Without Instructions: The Embryo as a Self-Organizing Field. 229
Because life cannot be "restarted" like a digital system. 231
From physics to consciousness: the conceptual journey of information. 233
Who is watching whom? The observer between the laboratory and lived experience. 235
When information changes the way we know. 237
Because life does not function like a computer. 240
Because an algorithm doesn't know it's alive. 242
From the bug to change: two opposing fates of error. 244
The Incomplete Code: Epigenetics, Space, and Time in Field Biology. 247
Memory without an archive: how cells remember who they are.... 249
Aging as an accumulation of biological memories. 252
The immune system does not forget. 255
The memory of living beings, from the cytoplasm to the forests. .. 258
Stones That Remember: When Matter Preserves Its Own History. 260
Philosophical Implications of Biological Space. 263

Embodied biography: inhabiting a body, becoming someone. 265
Technology and the Redefinition of Vital Places. 267
The Participating Gaze: Why Science, Too, Has a Place. 269
When information becomes an organism.........................271
The Eye and the Field: Seeing Beauty, Sensing Life. 273
VII. When Tradition Meets Science. 276
The word "energy": from a physical quantity to a metaphor for life. 279
From popular science to pop culture: the media's portrayal of energy. 281
From the page to the metaverse: how the narrative of life's energy will change. 283
Qi, prana, life force. 286
The knowledge of the hands and the breath: yoga, tai chi, and martial arts as lived biology. 288
From the temple to the supermarket: the ambiguous fate of vital energy...290
Not magic, but experience: because vital energy has endured through the centuries. 293
Counting Is Not Enough: Mathematical Bias and the Shadow Cast on Vital Energy. 295
PART FOUR. Life, Health, and Disharmony. 298
VIII. Health as Coherence,...301
Health is being in sync: the invisible orchestra of biological rhythms. 303
From the organism to the ecosystem: the same grammar of coherence. 304
Healing is not fixing: the living lexicon of health. 305
Health does not stand still: the biological art of remaining oneself while changing...308
Field resilience. Adaptation and self-repair. 310

What Life Remembers: Memory, Identity, and Biological Continuity. 311
Medical and ecological implications....... 314
One health, many living beings: when the biological realm transcends species. 316
The Invisible Guardians of the Field: When Health Speaks First to Shepherds. 318
Living Metropolises: When Built Space Enters the Body....... 320
IX. Illness as a Loss of Coherence....... 323
When Cells No Longer Understand Each Other: The Silence That Precedes Disease....... 326
Illness does not simply happen: it takes shape....... 329
Limitations and Possibilities of the Holistic View....... 331
Thinking in terms of fields, without omnipotence: what biology can learn from the planet....... 333
PART FIVE. Beyond Reductionism....... 336
X. Life as a Cognitive Process....... 339
When information is life, not computation....... 340
To live is to be ahead of the world. 342
The brain is not enough: where does cognition truly reside? 343
From the Heart to the Brain: A Western History of Localized Thought. 345
We also think with our gut: cognition beyond the brain....... 347
Knowing without knowing: the silent intelligence of life.350
Geese, spinarelli, and instant decisions: cognition without thought. 352
Measuring to Live: How Organisms Give Meaning to the World....... 355
Danger is sensed before it is calculated....... 357
Plants That Decide: How Life Measures Risk Without a Brain. 358

Making mistakes to stay alive: the prudent intelligence of life.......361
Life as a measuring tool: when science learns from organisms. 363
To measure is already to choose: the silent responsibility of the living. 365
Toward a Biology of Consciousness.................................368
Consciousness Put in Parentheses: How Biology Lost It and Is Rediscovering It. 370

XI—Toward a New Biology. 372

When science stops dismantling and begins to listen. 374
What changes for humanity when life is a field?.............377
Science and Metaphysics Reconciled. 380
Beyond explanation, before mystery: life as an event that speaks. .382

Conclusion. Life as an Event of Meaning in the Universe. 384

1 Bibliography. 387
Studies and Research 389
The Author 394

Introduction .

What is a living organism, really? An extremely complex biochemical machine, as twentieth-century biology has taught us, or something more: a structure of meaning, an information system capable of maintaining form, identity, and purpose over time? Field Biology arises from this radical question and takes the reader on a journey through science, philosophy, and the new frontiers of biological thought.

In recent decades, advances in molecular genetics and biochemistry have yielded extraordinary results, but they have also highlighted the limitations of a purely reductionist paradigm. Explaining life as the sum of chemical reactions or as the simple execution of a genetic code is no longer sufficient to account for fundamental phenomena: form, the coherence of the organism, self-organization, the capacity for creative adaptation, and, ultimately, biological meaning.

This book proposes a shift in perspective: viewing organisms as stabilized informational structures, supported by fields of coherence that precede and organize their individual material parts. This is not an escape into vagueness or symbolism, but a rigorous attempt to integrate contributions from the physics of complex systems, theoretical biology, neuroscience, and a critical reinterpretation of vitalist traditions in light of contemporary knowledge.

From the crisis of the classical biochemical paradigm to the unsolved problem of form in evolutionary biology, the text shows how life can be understood more effectively if conceived as a phenomenon of non-local coherence, in which the whole is not the result of the parts, but the condition of their possibility. From this perspective, biological fields are not merely evocative metaphors, but conceptual tools for describing how information takes shape, how energy becomes organization, and how the organism maintains its dynamic identity.

This vision also has profound implications for the way we understand health and disease. Health emerges as a state of global coherence, while pathology appears as a fragmentation of the field—a loss of informational harmony even before it is a mechanical failure. At the same time, life is reinterpreted as a diffuse cognitive process: every organism, by the very fact of its existence, perceives, evaluates, and responds to the world.

Field biology opens up a space for responsible reflection, in which science and metaphysics resume their dialogue without becoming conflated. It is an invitation to rethink life not only as a natural phenomenon, but as an event imbued with meaning in the universe, and biological knowledge as a form of understanding that inevitably involves cultural and ethical choices as well.

Introduction. A New Conceptual Grammar.

Why life can no longer be conceived of as matter in motion.

For over a century, biology has spoken the language of classical physics: matter, collisions, reactions, causal chains. A powerful language that has produced extraordinary results, but one that today shows increasingly evident cracks. Not because data is lacking—on the contrary—but because that data seems to say more than traditional categories are capable of grasping. This introduction stems from this: from the increasingly evident sense that life can no longer be conceived of as mere matter in motion, as a sum of chemical reactions orchestrated by chance and filtered by necessity.

The purpose of this introduction is not to demonstrate a new biological theory, nor to propose alternative experiments. Its task is more radical and, at the same time, more restrained: to lay the conceptual groundwork for the book, to clarify why it is necessary to change the grammar even before changing the model. The following pages invite the reader to momentarily set aside the automatic tendency to equate the living with the mechanical, in order to explore the idea that life is first and foremost a phenomenon of coherence, of field, of information taking shape.

In recent decades, molecular biology, the theory of complex systems, epigenetics, and certain convergences with contemporary physics have produced a silent paradox: we know more and more about the elementary components of life, yet we struggle to explain why the organism, as a whole, behaves as a unit endowed with stability, plasticity, and meaning. The problem, as Ludwig von Bertalanffy— , and founder of systems theory—already observed, lies not in the facts but in our perspective: *"Living systems are not aggregates of parts, but organized wholes."* A statement that today sounds less philosophical and more practical than it did in the mid-twentieth century.

This is where what we might call the silent crisis of modern biology emerges. Silent because it does not manifest as experimental failure, but as interpretive discomfort. Phenomena such as morphogenesis, the global regulation of the organism, and the surprising robustness of living systems in the face of disturbances transcend linear causality without thereby violating any natural law. As the biologist and Nobel laureate Albert Szent-Györgyi wrote, *"Life is a property of organization, not of molecules."* A simple, almost obvious observation, yet one that is deeply destabilizing for a biology accustomed to seeking ultimate causes in elementary building blocks.

When data outstrip theories, it is not uncommon for science to react by narrowing its focus, refining its tools, and accumulating further details. This book takes a different path: it asks whether it is not the conceptual framework that needs to be expanded. The idea of a biological field, which will be developed in the following sections, does not arise as a suggestive metaphor, but as an attempt to conceive of the organism as a stabilized informational structure, in which the whole precedes and guides the parts. In this sense, the reference to information should not be understood in an abstract or digital sense, but as an embodied principle, capable of generating form, function, and continuity over time.

A similar insight had been intuited, in another context, by the physicist David Bohm when he wrote that *"in an implicate order, the whole is present in every part."* Transposed into the biological domain, this insight suggests that living beings are not simply constructed from the bottom up, but are sustained from above by an order of coherence that does not coincide with any single material component. This is not a matter of abandoning biochemistry, but of recognizing its explanatory limitations when it is used as a total ontology.

From mechanism to meaning: this is the shift that this introduction aims to initiate. Speaking of meaning in biology does not amount to introducing naive finalism, but rather to recognizing that every organism constantly makes distinctions, maintains relationships, and preserves a form. In this sense, Francisco Varela's statement still resonates today: *"Living is a process of knowledge.*" A statement that opens the door to a biology capable of engaging in dialogue not only with physics but also with epistemology and the philosophy of nature, without losing its rigor.

This introduction is, therefore, a threshold. It offers not solutions, but guidance. It does not answer all the questions, but it clarifies why the questions themselves need to be reformulated. In the following chapters, the crisis of the classical paradigm, the notion of coherence, the idea of a field, and the implications for health and illness will be explored in greater depth. Here, the reader is simply invited to consider a possibility: that life is not merely what happens in matter, but the way in which matter is continually shaped, held together, and given meaning.

The words with which we conceive of life.

Every science is born twice: once in the laboratories and once in the language it uses to describe itself. Biology is no exception. Since its modern origins, it has borrowed words, images, and metaphors from physics and engineering, finding in them extraordinary explanatory power. Talking about "mechanisms," "molecular gears," and "genetic programs" has made it possible to make living things intelligible—to break them down, analyze them, and manipulate them. But today, just as biology has access to an unprecedented amount of data, those same metaphors are beginning to reveal their opaque side: rather than clarifying, they sometimes seem to obscure.

We need to focus on a point that is often overlooked: the problem is not just what we know about life, but how we describe it. Scientific language is not merely a neutral vehicle for information; it directs our gaze, suggests causal links, and determines what appears plausible and what is ruled out a priori. As the philosopher of science Mary Hesse noted, *"metaphors are not mere embellishments of scientific thought, but cognitive tools."* Tools that, precisely because they are powerful, can become invisible constraints.

The mechanistic metaphor has accompanied biology with almost perfect consistency. From the seventeenth century onward, the organism has been conceived as a sophisticated machine, the body as an apparatus, and the cell as a chemical factory. In the twentieth century, with the advent of molecular genetics, the vocabulary was updated without truly changing its underlying framework: DNA became a "code," genes "instructions," and the organism the execution of a "program." In 1961, when François Jacob spoke of *the "genetic program,"* the expression seemed illuminating; today, decades later, many biologists recognize that this metaphor risks suggesting a linear determinism that the actual data do not support.

The critical issue arises when language begins to precede observation, rather than follow it. Phenomena such as phenotypic plasticity, epigenetic regulation, or the self-organization of living systems struggle to find a place in a vocabulary designed to describe chains of command and instructional sequences. The British biologist Denis Noble observed, not without a certain irony, that *"genes are not commands, but resources."* Changing a word, in this case, means changing the very image of the organism: no longer a passive executor, but a dynamic system that integrates signals, contexts, and global constraints.

It is within this gap between language and phenomena that the proposal of this book finds its place. Speaking of fields, coherence, and embodied information does not amount to introducing vague or evocative terms, but rather to attempting a recalibration of scientific discourse . After all, the history of science shows that turning points often coincide with profound lexical shifts. When Michael Faraday introduced the concept of a field into physics, he was not adding a mysterious entity but providing a new way of expressing what the equations had already hinted at. Similarly, in biology, the use of less mechanistic language aims to make articulable what the data reveal but what the old metaphors can no longer contain.

This reflection does not imply a rejection of scientific precision, nor a flight toward poetic or symbolic language. On the contrary, as Niels Bohr emphasized, *"we are suspended in language,"* and precisely for this reason we must use it with rigor and awareness. The problem is not using metaphors, but forgetting that they are metaphors. When an organism is described exclusively as a machine, it becomes difficult to conceive of it as a unit endowed with history, memory, and dynamic stability. When life is reduced to matter in motion, what is lost is not the detail, but the overall meaning.

This paragraph therefore aims to offer the reader a framework for understanding: before delving into the themes of the crisis of the classical paradigm, biological coherence, and the fields of life, it is necessary to examine the vocabulary used to address these themes. Because, as the linguist Benjamin Lee Whorf wrote, *"we analyze nature along the lines drawn by our language."* Rethinking biology inevitably means also rethinking the words we use to describe it.

From physical fields to fields of life.

There is a subtle yet increasingly recognizable trend running through many contemporary scientific disciplines: the sense that the most interesting phenomena can no longer be described as mere chains of local causes and effects. In physics, neuroscience, ecology, and even in some branches of theoretical economics, words such as "field," "coherence," "emergence," and "global organization" have entered the lexicon not as a fad, but out of necessity. Biology, in this sense, is not an isolated exception, but rather the latest link in a broader transformation of the way we think about nature.

The best-known example is that of twentieth-century physics. When, between the 1920s and 1930s, quantum mechanics and field electrodynamics became established, it became clear that certain properties do not belong to individual objects but to the system as a whole. In Göttingen and Copenhagen—places that today seem almost mythical in the history of science—scientists debated the fact that a particle could no longer be conceived of as an isolated point, but rather as a node in a network of relationships. Without going into technical details, what matters is the shift in perspective: reality is not a sum of parts, but a coherent configuration.

Something similar is happening today in neuroscience. Beginning in the 1990s, with the development of *neuroimaging* techniques, it became apparent that functions such as perception or consciousness do not correspond to a single "center" in the brain, but rather to dynamic patterns of synchronization between distant areas. Neuroscientist Francisco Varela spoke of neural coherence to describe these transient yet highly organized states. Not a central command, but a temporary harmony, similar (as he said in a lecture in Paris in 1998) to a *jazz ensemble* rather than a classical orchestra.

Ecology also offers illuminating examples. Ecosystems are now studied as fields of nonlinear interactions, where stability does not depend on immutability but on the ability to absorb disturbances while maintaining an overall structure. C.S. Holling's work on ecological resilience, developed in Canada and Scandinavia in the 1970s, showed that a living system can change continuously without losing its identity. This idea resonates deeply with the biological notion of the organism as a stabilized informational structure.

This brief comparative overview is not intended to import models from one science to another, nor to force analogies. Rather, it serves to situate field biology within a broader cultural and scientific context, in which concepts such as coherence and field emerge spontaneously whenever one encounters complex, living systems. The feeling, shared by many researchers in different fields, is that reductionist language works well for individual components but loses its effectiveness when it comes to understanding the behavior of the whole. Biology today seems to be standing precisely at this threshold.

Biology that works but doesn't understand.

If one looks at contemporary biology from the outside, the idea of a "crisis" may seem almost paradoxical. Never before have we had such sophisticated tools, vast databases, and technologies capable of reading, rewriting, and visualizing vital processes with a precision that was unthinkable just a few decades ago. Yet, within the scientific community itself, a less triumphant sentiment is spreading: that of a discipline advancing rapidly, but without knowing exactly toward what conception of life it is heading.

It is a silent crisis because it produces no spectacular ruptures or open conflicts between schools of thought. There is no experiment that "doesn't work," nor any sensational anomaly that forces us to rewrite the textbooks overnight. The unease emerges elsewhere: in the way results are interpreted, categorized, and presented. As science historian Evelyn Fox Keller observed as early as the 1990s, the problem is not a lack of explanations, but an excess of partial explanations that do not come together to form a coherent whole.

An anecdote that has long circulated in the hallways of *the "Laboratory of Molecular* Biology" in Cambridge tells that Sydney Brenner, one of the fathers of molecular biology, loved to repeat—with a touch of provocation—*"We know everything about genes, but we still don't know what an organism is.*" The statement, attributed to an informal conversation in the early 2000s, was not an admission of ignorance, but a lucid assessment. Biology excels at analyzing components, but struggles to explain how these components come together to form a st r living entity capable of sustaining itself, adapting, and recognizing itself as a unit.

This interpretive challenge arises every time we attempt to link different levels of description. Genomic maps, regulatory networks, and metabolic fluxes provide extremely detailed information, but the transition from "how a part works" to "why the organism is this way and not another" remains problematic. Not because data is lacking, but because the dominant conceptual framework tends to treat the organism as the emergent result of local processes, without questioning the overall stability of the form.

This disconnect is also evident in scientific education. At many European and North American universities, including Heidelberg and Stanford, there has been discussion for years about students' difficulty in connecting molecular biology, physiology, and morphology. The disciplines coexist, but they speak slightly different languages. The result is fragmented knowledge—highly competent on a technical level, but hesitant when it comes to articulating a unified conception of life.

The crisis, therefore, does not concern the validity of the experiments, but rather the interpretive framework within which they are situated. Continuing to accumulate data without revising conceptual categories risks producing what the philosopher of science Thomas Kuhn would have called an increasingly efficient and increasingly unquestioning *"normal* science." In this sense, modern biology resembles an endless library in which the number of books increases, but the conceptual catalog remains static.

Speaking of an interpretive crisis does not mean calling for a change in method or a rebellion against experimental rigor. It means recognizing that the question "What is an organism?" cannot be resolved simply by improving the resolution of our instruments. It requires a rethinking of theoretical frameworks and greater attention to the principles of organization, continuity, and informational stability that hold living things together. It is at this subtle yet decisive threshold that the proposal for a biology of fields is situated.

Seeing is never neutral: images determine what life is.

Every scientific era views life through the images it is capable of producing. This is not an aesthetic detail, but a profound epistemological choice. When, in the second half of the nineteenth century, Rudolf Virchow asserted in Berlin that *"omnis cellula e* cellula" (*every cell arises from another cell*), he was not merely stating a biological principle: he was establishing a vision of the living shaped by the image under the optical microscope—the cell as a discrete, delimited, self-sufficient unit.

Since then, biology has continued to refine its visual tools—electron microscopy, fluorescence, molecular imaging—without, however, sufficiently questioning what these images implicitly suggest. Metabolic maps, genetic sequences visualized as linear codes, and cascade diagrams of cellular signaling pathways tell a specific story: that of a life that can be broken down, localized, and governed by specific causes. Yet, as the British biophysicist Brian Goodwin noted as early as the 1970s while observing the processes of embryonic morphogenesis, *"images explain less than they promise.*" They show us the where, rarely the how, and almost never the why.

A prime example is the famous "double helix" of DNA, made iconic by Watson and Crick's model in 1953 in Cambridge. A powerful image, capable of guiding decades of research, but also of reinforcing the idea that biological information resides in a static structure, readable like a text. By contrast, much less visual attention has been given to phenomena of epigenetic regulation, to the cellular electric fields studied by Lionel Jaffe at Berkeley, or to the dynamic l patterns observed in the 1990s in embryonic development laboratories in Kyoto and Tübingen.

Not because they were any less real, but because they were more difficult to "photograph." As the philosopher of science Ian Hacking noted, scientific images do not merely represent the world: they help establish what counts as real. In biology, this has meant prioritizing what can be isolated and stained, at the expense of what is relational, diffuse, and coherent. It is at this blind spot—silent yet decisive—that the perspective of the biological fields opens up: not against images, but beyond their power of simplification.

Laboratories Without a Horizon: The Fragmentation of Biological Knowledge.

There is one aspect of the crisis in contemporary biology that is rarely discussed, perhaps because it touches on raw nerves that extend beyond the laboratory: its social and institutional dimension. Today's biology does not suffer from a lack of data, nor from a shortage of technologies. On the contrary, it produces information in staggering quantities. Yet, as the sociologist of science Bruno Latour observed as early as the 1990s during his research on daily work in laboratories, knowledge never arises in the abstract: it is always embedded in networks of funding, careers, journals, calls for proposals, and productivity indicators. In this context, hyperspecialization is not merely an epistemological choice but a survival strategy. Studying a single protein, a narrowly defined metabolic pathway, or an isolated receptor is often more "fundable" than investigating the overall functioning of an organism.

A molecular biologist I met years ago at a European research center—let's call him M.—remarked with a touch of irony that in his department, *"everyone knows everything about almost nothing."* The joke hits on an essential point: the division of scientific work, driven by quantitative evaluation systems, tends to break down living systems into ever-smaller problems—

perfectly analyzable, yet increasingly difficult to reconcile within a unified framework. It is no coincidence that concepts such as "field," "coherence," or "global organization" are often perceived as vague or suspect: not because they lack foundation, but because they do not easily fit into *grant* applications or the criteria of a *high-impact-factor* journal.

As early as 1962, Thomas Kuhn, in his **The Structure of Scientific* Revolutions*, observed that paradigms do not change merely through the accumulation of anomalies, but when it becomes possible to imagine a different way of doing science. Today, in biology, many anomalies are plain for all to see: non-local regulations, long-distance synchronizations, and systemic resilience that are difficult to explain in purely biochemical terms. What is lacking are not the observations, but the institutional and cultural spaces to integrate them. Added to this is a paradoxical effect: the more technically powerful a discipline becomes, the more it tends to defend its own interpretive categories, because funding, reputations, and entire research schools depend on them. As Nobel laureate Peter Medawar noted in a 1974 lecture in London, *"science can become conservative precisely when it seems most advanced."*

In this scenario, speaking of organisms as stabilized informational structures does not amount to heresy, but rather to an attempt at recomposition. It is not a matter of denying the value of molecular analyses, but rather of recognizing that they describe fragments of a broader process, one that requires a language capable of spanning different levels. Field biology also emerged as a response to an implicitly sociological question: how to restore a sense of unity to a science that, given the way it is organized, risks losing sight of the big picture even as it accumulates an unprecedented wealth of detail.

Predicting is not the same as understanding: the paradox of contemporary biology.

One of the most surprising—and least discussed—features of contemporary biology is the growing gap between what we know how to do and what we know how to understand. We have become extraordinarily skilled at predicting and manipulating biological processes: we anticipate gene expression, modify a protein, block an inflammatory cascade, and guide a stem cell toward one fate rather than another. Yet, if we pause for a moment, the simplest question often remains unanswered: what, taken as a whole, is the organism we are controlling? This tension is not new. As early as 1917, in Vienna, Ludwig von Bertalanffy noted in his notebooks that *"the ability to predict does not coincide with an understanding of the organization."* A century later, the gap between the two seems to have widened.

A telling example comes from the recent history of pharmacology. In the 2000s, at various research centers from Boston to Basel, scientists refined their ability to design highly selective drugs capable of targeting a single molecular target with almost surgical precision. The experimental predictions held true: the receptor was inhibited, and the signal was blocked. And yet, in clinical trials, unexpected systemic effects multiplied. As one researcher involved in one of these programs put it, *"Everything was under control, except the organism."* Local prediction did not translate into global understanding. The problem was not technical, but conceptual: the organism responded as a coherent unit, not as a sum of independent parts.

This discrepancy recalls a classic distinction, dear to the German philosophical tradition, between Erklären and Verstehen: explaining a mechanism is not the same as understanding a meaning. In biology, causal explanation has produced extraordinary results, but it tends to lose sight of the organizational- l dimension—the one that gives meaning to local interactions. Norbert Wiener, the founder of cybernetics, noted in 1950 that *"a system can be predictable and yet incomprehensible if we ignore the principle that organizes it."* It is precisely here that the perspective of biological fields introduces a discontinuity: it does not deny prediction, but relativizes it, placing it within a framework in which the behavior of the parts acquires meaning only in reference to an informed whole.

Field biology suggests that understanding an organism does not merely mean anticipating its responses, but grasping its internal logic of stabilization—the implicit rules that maintain form, function, and identity over time. As the neurophysiologist Kurt Goldstein observed while working with brain-injured patients in Frankfurt in the 1930s, the organism does not simply "react" to stimuli: it reorganizes itself. Predicting a single experimental outcome may be useful; understanding the organization that makes that outcome possible is another undertaking—one that is slower and less immediately applicable, but decisive if we truly wish to speak of life as a meaningful phenomenon.

Rebellious Data: What Organisms Do That Theories Fail to Explain.

There comes a moment in the history of every science when the data begins to misbehave. Not because it is wrong, but because it no longer fits where it's supposed to. It continues to accumulate; it is reproducible, solid, often even spectacular—and yet it does not fit into the available theoretical frameworks. In biology, this moment is neither sudden nor explicit: it is silent, gradual, almost embarrassing. The observed phenomena work, but they don't "add up." It is like attending a concert where every musician plays correctly, yet the overall harmony defies any known score.

A classic example comes from morphogenesis, the study of how an organism takes shape. In 1952, in an office at the University of Manchester, Alan Turing—the same mathematician who had helped decipher Enigma—published a surprising article in **Philosophical Transactions of the Royal Society**. He demonstrated that simple chemical reactions, when coupled with diffusion processes, could generate complex patterns: stripes, spots, symmetries. The model worked, and it is still used today. But even then, it was clear that explaining how a form emerges was not the same as explaining why that form remains stable, recognizable, and capable of regenerating if damaged. The data revealed a regularity that went beyond the local cause-and-effect chain.

Even more baffling is the case of regeneration. Anyone who has observed a planarian—a small, flat, worm that has been studied since the 19th century—knows that it can reconstruct an entire organism from a tiny fragment. When cut into several pieces, it does not produce a random mass of cells, but rather new, complete individuals, with head and tail in the right places. During experiments conducted in the 1920s and 1930s by Thomas Hunt Morgan in Woods Hole, Massachusetts, the phenomenon was already well documented. The problem was not empirical; it was theoretical. No linear chain of local instructions seemed sufficient to explain how the cells "knew" where to position themselves. The data pointed to a global control mechanism, but the available terminology spoke only of localized reactions.

Similar phenomena emerge in contemporary cell biology. So-called *"quorum sensing*," first observed in the 1970s in marine bacteria studied in Honolulu, shows how unicellular populations modify their collective behavior based on density, activating genes only when a critical threshold is reached. No single cell, taken in isolation, contains the cause of the change. It is the system as a whole that produces a new state. Causality is no longer linear but distributed: not "A causes B," but rather "a configuration makes an event possible."

Even genetics, long considered the bastion of deterministic explanation, has been confronted with an excess of data. Barbara McClintock, working on corn in the 1940s and 1950s at Cold Spring Harbor, observed that the genome is not a static text, but a dynamic system capable of reorganizing itself in response to stress and context. She spoke of a "responsive genome," a term that earned her decades of misunderstanding before her 1983 Nobel Prize. The data were there, but the theories were not ready to accommodate them.

These examples, diverse in scale and context, converge on a crucial point: living organisms exhibit emergent properties that cannot be reduced to linear chains of causes. Field biology arises precisely in this space of excess, where the facts demand a different grammar. Not to abandon rigor, but to recognize that the organism operates as a stabilized informational structure, capable of maintaining coherence through flows, perturbations, and changes. The data, after all, are not betraying science: they are signaling that it is time to expand its language.

Emergence without mystery.

For many biologists, uttering the word "emergence" has long been an ambiguous gesture, almost an elegant surrender. As if saying that a property "emerges" meant admitting that we cannot truly explain how and why it happens. Yet, upon careful observation of the recent history of the life sciences, it becomes increasingly clear that emergence is not a black hole of knowledge, but one of its most refined tools. It doesn't eliminate the explanation—it shifts it to a different level.

An anecdote often recounted in the hallways of the Santa Fe Institute in New Mexico helps clarify the point. In the early 1990s, during an informal seminar, a young researcher demonstrated how very simple computational models could generate unpredictable collective behaviors. At the end of the presentation, a senior physicist commented, *"We haven't lost causality; we've distributed it.*" That sentence already captured the heart of emergence: not the absence of causes, but their organization into configurations that produce novelty.

In biology, emergence becomes necessary when the observed properties do not belong to any of the parts taken individually. An organism's consciousness, its capacity for creative adaptation, and the stability of a form over time cannot be localized to a single molecule, gene, or cell. They are properties of the system as a whole. Ludwig von Bertalanffy, the founder of systems theory, wrote as much as early as 1968 in his *General System Theory*: *"The whole is more than the sum of its parts—not by magic, but by organization.*" A y statement that, decades later, sounds less philosophical and more practical.

A particularly instructive example comes from developmental biology. In the 1980s, in the EMBL laboratories in Heidelberg, researchers observed that embryos subjected to significant disturbances still managed to achieve a correct final structure. Not because every error was locally corrected, but because the organism seemed to "tend" toward an overall form. Speaking of emergence, in this case, did not mean introducing an occult principle, but rather recognizing that the explanation lay in collective dynamics, not in individual steps.

The misunderstanding often stems from an implicit comparison with the classical mechanistic ideal, in which explaining is equivalent to breaking things down. But contemporary science—from the thermodynamics of systems far from equilibrium to network theory—has shown that understanding can also mean following the emergence of patterns, constraints, and regularities that were not "written" in any isolated part. The vortices studied by Ilya Prigogine in the 1970s, or the dissipative structures observed in real chemical systems, were not "mysterious": they were emergent, and precisely for this reason intelligible on a new level.

From this perspective, emergence becomes a fundamental key to conceiving of the organism as a stabilized informational structure—not a complicated machine, but a process that generates and maintains biological meaning. Far from representing a step backward in terms of explanation, emergence marks a step forward: it acknowledges that life cannot be exhausted by a single descriptive scale and that true understanding arises in the transition between levels. The mystery, if anything, is not that order emerges, but that for so long we thought we could explain it without taking this into account.

Between Expectations and Thresholds: Biology and Time That Doesn't Flow Linearly.

There is an unspoken assumption that underlies much of modern biology: the idea that time flows uniformly, reversibly, and neutrally. A laboratory-style time, marked by stopwatches and Cartesian graphs, in which a cause precedes an effect with almost clockwork precision. Yet anyone who has truly observed a living organism—a plant that "decides" when to bloom, an immune system that reacts too late or too early, a scar that remains as a mark—knows that biological time is something else entirely. It is irregular, layered, and often asymmetrical. In a word: nonlinear.

In the 1950s, the British biologist Conrad Hal Waddington, working in Edinburgh, introduced the image of the epigenetic landscape: a ball rolling across a surface of valleys and ridges, where the path depends not only on the initial push but also on the history already traversed. It was not just an elegant metaphor: it was a way of saying that, in biology, the past remains operative. Cells "remember" what has happened, and this memory influences what may happen. Time, here, is not an external variable: it is embedded in the form.

A less theoretical example comes from clinical observation. In the 1990s, in some European oncology departments, it was noted that seemingly similar tumors, treated with identical protocols, showed radically different responses. It was not merely a genetic issue. Over time, it became clear that delays, thresholds that were either exceeded or not reached, and the temporal sequences of therapies all mattered. A drug given "too early" could be ineffective; one administered "too late" could be devastating. The biological system did not react proportionally, but in leaps, discontinuities, and points of no return.

This introduces another decisive characteristic of living temporality: irreversibility. An organism never returns exactly to its previous state. Even when it heals, it remains different. As the French physiologist Henri Atlan wrote in the late 1970s, *"biological order is constructed by consuming possibilities."* Every temporal choice opens some paths and closes others. Life is not a tape that can be rewound, but a story that is written as it unfolds.

It is not surprising, then, that many linear models struggle to describe phenomena such as latency, slow adaptation, or delayed responses. Consider the endocrine system: a signal may be emitted now and produce significant effects only days or months later. Or consider the gut microbiota, which has been studied intensively since 2010, where minimal and initially invisible changes can, over time, reorganize entire physiological balances. Here, time acts as a silent amplifier, not merely as a coordinate.

This complex temporality compels us to rethink the organism as an informational structure stabilized over time. Stabilized, but not fixed. Biological information is not an instantaneous message: it is a process that accumulates, transforms, and is sometimes reactivated. Just as in certain historic neighborhoods of cities like Lisbon or Naples, where layers from different eras coexist in the same space, so too do different times coexist within organisms: evolutionary, ontogenetic, and physiological. Understanding life means learning to read them together, without reducing them to a single timeline.

When the experiment succeeds, but the theory remains silent.

In laboratories around the world, a curious, almost embarrassing scene is playing out with increasing frequency. The experiment has succeeded; the data are clean, reproducible, and publishable in leading journals. Yet, when it comes to explaining why that phenomenon occurs, the language becomes cautious, interpretations diverge, and conclusions remain unresolved. It is as if contemporary bi l has become extraordinarily skilled at producing correct answers to poorly phrased questions.

A prime example comes from *Tufts University*, in Michael Levin's laboratory, where bioelectric patterns in organism development have been studied for over a decade. By modifying extremely weak electrical gradients in the cells of frog embryos (*Xenopus laevis*), the team has obtained results that seem straight out of a science fiction story: eyes forming in abnormal positions, limbs regenerating where they shouldn't, tissues that "know" what to become even when the genetic context has been disrupted. The experiments are robust, replicated, and filmed. But what really drives form? A gene? A local signal? Or something more like a distributed information field, capable of coordinating the whole?

Another example, less well-known to the general public but much discussed among experts, concerns so-called "biological robustness." In numerous developmental biology experiments conducted in Germany and Japan in the early 2000s, it was observed that embryos subjected to significant perturbations—temperature variations, cell removal, chemical alterations—still manage to develop surprisingly normal forms. Classical models speak of local molecular compensations. But when these compensations are too numerous, too coordinated, and too reliable, the idea of a simple sum of adjustments begins to falter.

Then there is the now-famous case of photosynthesis. In 2007, a group led by Graham Fleming published experimental data in *Nature* showing signs of quantum coherence in the photosynthetic complexes of algae and bacteria. The results were clear: energy did not "jump" randomly from one molecule to another, but seemed to explore multiple pathways simultaneously, as if the system knew in advance which one was the most efficient. The scientific community reacted with interest, but also with unease. The phenomenon was real; the interpretation, much less so. Was it an artifact? A borderline case? Or a clue that living systems— —utilize organizational principles that classical biology has never truly integrated?

In medicine, too, history repeats itself. In fecal microbiota transplants—performed with increasing success for certain intestinal disorders—clinical results can be excellent. Yet two patients treated with nearly identical protocols may show radically different outcomes. The transferred bacteria are the same, the conditions are controlled, and the data are accurate. What is missing is a theory capable of explaining how a complex living ecosystem reorganizes itself as a whole, and why this reorganization follows trajectories that cannot be predicted based on its individual components.

In all these cases, the problem is not the unreliability of the experiment. On the contrary: it is its excessive success. The data exceed the available conceptual categories. As the physicist and philosopher of science Peter Medawar noted as early as 1968, *"facts do not speak for themselves unless there is an adequate language to listen to them.*" Field biology arises precisely in this gap: where the organism no longer appears as a complicated machine, but as a stabilized informational structure, capable of maintaining form, function, and meaning even when its individual parts are altered.

When we control the outcome but lose the meaning.

One of the most celebrated triumphs of contemporary biology is its ability to produce effects. We switch off a gene, and a protein disappears. We activate a receptor, and a cell changes its behavior. We insert a sequence and obtain a measurable result, often reproducible with impressive precision. But a gap opens up between the effect obtained and the form understood—a gap that the rhetoric of scientific progress tends to downplay.

A telling example comes from the world of organoids—those tiny three-dimensional structures grown in the laboratory from stem cells. In 2009, in the laboratories of the *Hubrecht* Institute in Utrecht, the group led by Hans Clevers succeeded in growing small, functional "intestines" in vitro. The result was extraordinary: isolated cells that, without an explicit blueprint, spontaneously organized themselves into a structure featuring crypts, villi, and functional differentiation. The experimental prediction held true. But when it came to the crucial question—where does form come from?—the answers remained vague. There was no "architect" gene, nor a linear sequence of instructions. Rather, there was a collective dynamic that seemed to precede and guide individual molecular events.

Something similar happens with gene-editing techniques. CRISPR-Cas9, developed starting in 2012 thanks to the work of Jennifer Doudna and Emmanuelle Charpentier, allows for interventions with surgical precision. We can predict the effect of a mutation with a very high degree of certainty. Yet, in many cases, the organism reacts in unexpected ways: alternative pathways are activated, functions emerge elsewhere, and the overall structure reorganizes itself. As the British biologist Brian Goodwin observed as early as the 1970s, intervening on a single element does not equate to understanding the logic that holds the whole together. It is the difference between knowing how to achieve a result and knowing why that result is possible.

The history of developmental biology is littered with these interpretive short circuits. In the 1950s, at Woods Hole, Conrad Waddington introduced the concept of the "epigenetic landscape" to explain why genetically identical cells could follow different fates. It was a powerful metaphor, but also an implicit admission: the known mechanisms were not enough. Today we have infinitely more sophisticated tools, but the question remains surprisingly similar. Form is not simply the consequence of a causal chain; it is a global property that emerges from relationships distributed across space and time.

This distinction between effect and form is not merely theoretical. In regenerative medicine, for example, one can stimulate cell proliferation or the expression of growth factors, achieving temporary results. But reconstructing a functional structure—a tissue, an organ, a physiological balance—requires something more: an organizing principle that coordinates local processes into a coherent whole. This is where the notion of the biological field begins to take hold, not as an escape into vagueness, but as an attempt to name what the experiment reveals but fails to explain.

As the philosopher of science Georges Canguilhem wrote, life is not defined by its components, but by the normativity that organizes them. To predict an effect means to operate within an already established norm. To understand form means to ask how that norm is continually generated, maintained, and, at times, transformed. It is a subtle yet decisive shift: from biology as the engineering of results to biology as the science of meaning.

Restless Numbers: What Biological Data Do Not Reveal on Their Own.

There was a time when scientific data was accepted as an answer—a fixed point, a confirmation, sometimes a victory. In contemporary biology, this attitude survives more as a habit than as a genuine conviction. Data today is abundant, almost to the point of redundancy, but it rarely brings a discussion to a close. On the contrary, it opens it up—and often in unexpected directions.

This is clearly evident in the major research centers producing global biological atlases. In 2016, when the *"Human Cell* Atlas" project was unveiled in London, the stated goal was "to map all the cells in the human body ." An ambitious promise, supported by increasingly sophisticated single-cell sequencing technologies. But after just a few years, the researchers involved—from Cambridge, Stanford, and the EMBL in Heidelberg—found themselves facing a paradox: the more cells were classified, the more unstable the categories became. Cell types that blurred into one another, intermediate states, continuous transitions. The data did not provide a definitive answer to "what a cell is," but raised a deeper question: how is biological identity maintained amid constant change?

This shift is crucial. In the field of biology, data is not the ultimate building block of knowledge, but a signal. It indicates that there is an order, without, however, exhausting it. As was the case with Claude Bernard in mid-19th-century Paris: while studying the *"milieu intérieur,"* he realized that the stability of organisms did not stem from the absence of variations, but from their continuous regulation. His experiments provided precise measurements, but what really mattered was the question that emerged from them: what principle holds this apparent instability together?

Even today, much biological data works in the same way. Take, for example, the metabolic oscillations observed in yeast, which have been studied since the 1990s in laboratories in Gothenburg. The graphs show regular cycles, synchronized fluctuations, and collective behavior among cells. It is all measurable and repeatable. But the meaning is not contained in the numbers. The data asks to be interpreted as a trace of an invisible coordination, of a collective dynamic that cannot be reduced to the sum of individual processes.

In this sense, biological data resembles a philosophical question more than an engineering answer. It does not simply state *"this happens,"* but suggests *"how is it possible for this to happen?"* A subtle but decisive difference. As the physicist and philosopher Erwin Schrödinger noted in his lectures in Dublin in 1943, the facts of life are not strange because they violate known laws, but because they use them in a ly organized way. The data, then, is not the end of the investigation: it is the beginning of a reflection on organization.

Changing our perspective on data also means accepting a certain degree of incompleteness. In the hallways of *the Max Planck Institute* in Dresden, where tissue self-organization processes are studied, one sometimes hears young researchers say that *"the data are beautiful, but we don't yet know what they're asking of us."* It is a revealing statement. It points to a science that no longer seeks to dominate the phenomenon, but to listen to it.

From this perspective, field biology does not propose abandoning data, but rather restoring its depth. It views data not as definitive answers, but as guiding questions. Every measurement then becomes a window onto a broader order, every anomaly a clue, every regularity a disguised question. It is perhaps here that biology ceases to be merely a science of life and begins to become a science of the meaning of life.

Beyond the Gearwork: Why Organisms Are Not Well-Assembled Machines.

For over a century, biology has spoken the language of mechanism. A powerful, precise language, capable of producing extraordinary results: targeted drugs, sophisticated imaging techniques, and genetic interventions that were unthinkable just a few decades ago. Yet, beneath this surface of success, a silent crack has opened. Organisms continue to be described as devices that function, but ever more rarely as systems that convey meaning.

The shift from mechanism to meaning does not occur through a sudden theoretical rebellion, but through saturation. When every vital process is explained as a chain of efficient causes, a fundamental question remains unanswered: why this particular organization and not another? Why a stable form, a coordinated function, a contextual response that seems "sensible" given the situation? Mechanical language explains how something happens, but remains silent on why that thing makes sense for the organism.

An episode often recalled in systems biology labs at MIT concerns the first computational simulations of genetic networks in the early 2000s. The models worked: they produced oscillations, stable states, and transitions. But, as Stuart Kauffman recounted during an informal seminar in 2002, *"we couldn't tell whether the system was doing something or simply happening.*" It was an important admission: the behavior was correct, but the meaning was not. A biological semantics was missing.

This is where the field perspective comes into play. Speaking of a "field" does not mean invoking vague metaphors, but rather recognizing that biological organization precedes and guides individual events. Just as in a magnetic field, where iron filings reveal an invisible structure, so too in living systems do molecular processes "align" with an order that is not contained within any single isolated reaction. Information, in this context, is not a message traveling from one point to another, but a relationship that shapes overall behavior.

An illuminating example comes from developmental biology. In 1924, in Berlin, Hans Spemann demonstrated that by transplanting a small embryonic region—the famous "organizer"—it was possible to induce the formation of a second body axis. Spemann spoke of "morphogenetic fields," using terminology with great caution. He did not know how that effect occurred at the molecular level, but he had grasped something essential: biological information resides not only in the components, but in the configuration that makes them function as a whole.

This insight resurfaces today, using different tools, in the study of cellular and tissue networks. When an organism responds to an injury, environmental stress, or a nutritional change, it does not follow a rigid protocol. It interprets a situation. And to interpret means to assign significance, establish priorities, and modulate responses. It is an elementary language, devoid of reflective consciousness, but not devoid of meaning.

It is no coincidence that philosophers of science such as Georges Canguilhem insisted, as early as the 1950s, on the normative character of life. In a lecture at the Sorbonne in 1952, Canguilhem observed that an organism does not merely endure its environment: it evaluates it, distinguishing between what is favorable and what is unfavorable. This capacity for evaluation is not an accessory, but the very heart of the living. It is here that the mechanism reveals its conceptual limit.

The perspective of the field and embodied information thus allows for a shift in perspective. Life no longer appears as an object that functions, but as a process that— —means something in and of itself. Every organism becomes a stabilized informational structure: not a data archive, but a system that maintains its own identity by continuously interpreting signals, constraints, and possibilities. As the Chilean biologist Francisco Varela wrote in his 1987 research notebook, now preserved in Santiago: *"To live is to produce meaning, even when there is no one to observe it.*"

In this transition from mechanism to meaning, biology does not lose its rigor. Its grammar changes. And perhaps, precisely for this reason, it comes closer to what it seeks to understand.

The hidden language of organisms.

There is a silent paradox running through contemporary biology. Never before have we had access to such an immense amount of data: sequenced genomes, mapped metabolic networks, ultra-high-resolution cellular images. Yet the more detail we have, the more life seems to elude a unified understanding. It's like observing a conversation while knowing the chemistry of ink perfectly, but ignoring the meaning of the sentences.

In recent decades, a question has begun to creep into the margins of biological research: what if organisms were not merely biochemical machines, but systems capable of producing, interpreting, and transforming signs? Not poetic metaphors, but signs in the strongest sense of the term. This is where biosemiotics comes into play—a field of study that emerged in the late 1980s and early 1990s, often far from the spotlight, yet increasingly difficult to ignore.

The Danish biologist Jesper Hoffmeyer liked to say that the first mistake of modern biology was to confuse cause with meaning. A chemical reaction can explain how something happens, but not what that event means for the organism experiencing it. In biosemiotics, the key concept— —is that of the sign: something that stands for something else to someone. In an organism, that "someone" is not a reflective consciousness, but the entire living system.

Consider a bacterium that alters its movement in response to a chemical change in its environment. From a classical perspective, it is a cascade of molecular reactions. From a semiotic perspective, that external molecule functions as a signal signifying "food" or "danger." It is not the molecule itself that guides behavior, but the meaning it takes on within the history and organization of the living being.

In 2001, during a conference in Tartu—an Estonian city inextricably linked to Juri Lotman's semiotic tradition—a young researcher described how the same chemical substance could elicit opposite responses in genetically similar organisms raised in different environments. *"They don't react,"* he said, *"they interpret."* The room fell silent for a few seconds: it was not a provocation, but an empirical observation that called for new conceptual categories.

Biosemiotics does not deny biochemistry, just as linguistics does not deny the physics of sound waves. It transcends it through integration. Every organism, from the simplest to the most complex, lives immersed in a network of signs that it itself helps to generate. Hormones, electrical gradients, spatial configurations of tissues: all of this is not merely "matter in motion," but information that holds significance for the system.

A fascinating example comes from botany. In the 2000s, in an experimental greenhouse near Florence, a team observed that certain plants modified their root development not based on the physical presence of other roots, but in response to specific molecules released into the soil. Those molecules were neither nutrients nor toxins: they were signals. The plant "read" its environment and made morphological decisions consistent with that reading.

Here, the sign is not an arbitrary symbol, but a stable relationship between form, function, and context. This is what makes it possible to speak of organisms as stabilized informational structures: entities that maintain their identity not only through energy and matter, but through the continuity of the biological meanings they are able to recognize.

Unlike human languages, the language of life is not codified in dictionaries nor governed by explicit syntax. It is an immanent grammar that emerges from the evolutionary history and internal organization of organisms. As early as the first decades of the twentieth century, the biologist and philosopher Jakob von Uexküll spoke of *the Umwelt*: the meaningful world specific to each species. Not an objective environment, but a universe of relevant signs.

Re-examined today, Uexküll appears surprisingly relevant. His work suggests that biological reality is not a neutral stage, but a fabric of interpreted relationships. Every organism lives in a world that makes sense to it, and this capacity for meaning is what we call life.

Accepting this perspective means shifting our focus: asking not only what components an organism is made of, but what signs it is capable of recognizing, what information it can embody, and what consistencies it manages to maintain over time. It is a shift in language even before it is a shift in theory.

Speaking of signs in biology is neither a return to vitalism nor a surrender to mysticism. It is a rigorous and necessary attempt to fill a conceptual gap. As the philosopher of science Evelyn Fox Keller observed in a lecture in Berlin in 2012, "*We have learned to describe the components of life, but not yet its discourse.*"

This introduction does not claim to provide definitive answers. Rather, it aims to open a path: to suggest that life, before it can be explained, must be understood as a phenomenon of signification. An event in which matter becomes capable of saying something to itself.

Information Without a Sender

We are accustomed to thinking of information as something that someone sends to someone else. A sender, a message, a recipient. This is the model of communication we've inherited from technology, from the nineteenth-century telegraph to digital algorithms. But what happens if we try to apply this framework to life? Who "sends" the information to a cell that differentiates, to an embryo that takes shape, to an organism that adapts to a constantly changing environment?

Contemporary biology is increasingly confronted with informational phenomena lacking a recognizable sender. There is no command center, no hidden conductor. Yet the information is there: it circulates, stabilizes, and produces form and behavior. It is information that is not sent, but emerges.

In the 1950s, the mathematician Claude Shannon developed a highly powerful theory of information, but one that was deliberately silent on meaning. It worked very well for telephone cables, but much less so for living systems. In an organism, information is not an abstract quantity that travels from A to B, but an event that occurs "when a difference makes a difference," to use an expression made famous by the anthropologist Gregory Bateson.

Bateson, who loved to say that he had learned more by observing dolphins than by reading textbooks, insisted on a crucial point: in living systems, information arises from relationships. It is not contained in an isolated signal, but in the way a system is structured to respond to certain variations and not to others. There is no need for intention; all that is needed is an internal coherence that makes that change relevant.

A fascinating example comes from studies of *Physarum polycephalum*, a slime mold that has been studied since the 1970s in Japanese and American laboratories. This organism, which lacks a nervous system, is capable of "solving" mazes and optimizing paths to food. No neuron sends instructions; no central hub makes decisions. Information emerges from the dynamics of the body itself, from the continuous dialogue between form, movement, and the environment.

Talking about information without a sender means shifting the focus from the signal to the interface. Information is neither entirely within the organism nor entirely outside it: it arises in the space of contact. In the 1960s, psychologist James J. Gibson introduced the concept of "affordance" to describe the possibilities for action that the environment offers an organism. A surface "tells" only those with a certain body, a certain scale, and a certain evolutionary history that it can be walked upon.

In this sense, the environment does not send messages: it proposes configurations. It is the organism, with its stabilized information structure, that renders those configurations meaningful. Information does not precede the relationship, but accompanies it. It is a necessary byproduct of being-in-relation.

A researcher recounted realizing this while observing bacterial colonies under thermal stress. There was no new chemical signal, no identifiable molecular "alarm." Yet, collective behavior changed in a coordinated manner. The information was not contained in a specific molecule, but in the overall change in the physical context and in the system's ability to interpret it as such.

From this perspective, information is not content that is transmitted, but an effective configuration. Living systems do not receive instructions: they organize themselves in such a way that certain events become informative. It is a subtle but decisive difference. As theoretical biologist Francisco Varela observed in a lecture in Paris in 1996, *"cognition is not the representation of a given world, but the emergence of a workable world."*

This is where the idea of a biological field comes in. The field does not transmit messages, but stabilizes relationships. It is what makes it possible for information to emerge without being sent, for a form to be maintained without an external design. In this sense, organisms are informational structures not because they contain data, but because they make the world informable.

Accepting the idea of information without a sender also means letting go of a certain reassuring image of life as a system controlled at every moment from above. Vital processes do not carry out orders: they respond to constraints, opportunities, and resonances. The biologist and chemist Ilya Prigogine, a 1977 Nobel laureate, spoke of "order through fluctuations": structures that arise not in spite of disorder, but thanks to it.

In this light, life appears less like a machine and more like a diffuse interpretive process, without a central interpreter. A process in which information has no author, but does have meaning. And perhaps this is precisely what makes living things so difficult to reduce to mere matter in motion.

Technology and Misunderstandings of the Living.

Every era understands nature through the conceptual tools at its disposal. In the seventeenth century, the universe was a giant clock; in the nineteenth century, a steam engine; today, inevitably, a computer. Biology was no exception. Indeed, it is perhaps the discipline that has most deeply absorbed the technological metaphors of its time. Speaking of a "genetic code," "developmental programs," and "transcription errors" has had enormous explanatory power. But, as often happens with successful metaphors, their success has ultimately obscured their limitations.

The idea that life functions like software running on molecular hardware arose in a specific historical context. We are in the 1940s and 1950s, in Princeton, Cambridge, and Los Alamos. It was the era of cybernetics, the first electronic computers, and the Macy conferences. It is not surprising that, just a few years later, Francis Crick could speak of DNA as a "code" and Jacques Monod, in * *, could describe the organism as a chemical machine governed by instructions. The computer metaphor finally seemed to offer a rigorous language for what had until then appeared mysterious.

Yet even then, some people sensed unease. In 1967, in Santa Fe, biologist Conrad Hal Waddington—the same man who had introduced the image of the "epigenetic landscape"—wryly observed that no embryo had ever read its own genome. A joke, certainly, but a revealing one: the problem is not DNA, but rather the idea that biological information functions as an instruction manual.

Computational metaphors worked because they enabled operational science: sequencing, comparing, modifying. In the 1990s, during the *Human Genome Project*, laboratories in Bethesda and Cambridge spoke openly of genetic "debugging." The imagery was that of engineering: if there's an error, you correct it. But while software remains identical to itself regardless of the computer on which it runs, a gene outside its biological context loses much of its meaning.

An anecdote told by a French geneticist is illuminating. In the early 2000s, during a gene transfer experiment on model plants, the "right" gene produced radically different effects depending on the tissue, the stage of development, and even the lighting conditions in the greenhouse. "*It was the same code*," she said, "*but it seemed to speak different languages*." The problem lay not in the sequence, but in the idea that the sequence was the message.

Herein lies the central misunderstanding: confusing biological information with technical information. In technological systems, meaning is external: we decide what a bit represents. In living systems, however, meaning is internal to the process itself. It is not assigned, but embodied.

The image of a program suggests a linear sequence, a beginning and an end, and top-down control. But life does not "execute" anything. It organizes itself, reorganizes itself, deviates, and explores. In 1981, evolutionary biologist Stephen Jay Gould described how evolution resembles a jazz improvisation more than a classical score. A musical metaphor—not a computational one—that better captures the temporal, situated, and relational dimension of living beings.

Thinking of an organism as a computer leads us to search for the "command center," the main file, the original instruction. But real organisms function through distribution, redundancy, and plasticity. Paradoxically, digital technology has helped us understand life only when it began to resemble it less: through neural networks, adaptive systems, and unsupervised learning. This is a sign that the problem isn't the use of metaphors, but taking them literally.

The point is not to abandon the language of information, but to free it from the imagery of the machine. In biology, information is not a text to be deciphered, but a dynamic to be inhabited. Reducing it to code means losing sight of what makes it effective: the context, the history, and the organism's overall form.

In 2005, during a seminar in Heidelberg, a young researcher provocatively asked whether an organism was more like a computer or a city. The answer was not unanimous, but many agreed on one point: a city does not function because someone possesses a complete map of it, but because millions of local interactions produce a recognizable order. Life, perhaps, is more urban than computational.

In this sense, the perspective of the biological sciences does not reject technology, but puts it into perspective. It reminds us that every conceptual tool illuminates one aspect while obscuring another. And that living things, when understood solely through the metaphors of the machines we build, inevitably end up being misunderstood.

From Calculation to Meaning: The New Unease in the Exact Sciences.

For much of the twentieth century, speaking of "meaning" in the hard sciences was considered, at best, a philosophical quirk; at worst, an unsavory deviation. Equations worked without meaning, models predicted without understanding, and this seemed to suffice. Yet something began to crack precisely where formalization was most rigorous—not due to a lack of data or computational power, but because of a more subtle unease: the growing impression that describing is not the same as understanding.

In physics, the warning signs came from afar. As early as the 1980s, in the corridors of CERN in Geneva, some theorists—including Carlo Rovelli, then a young researcher—began to question whether the fundamental laws described not "things," but relationships. Loop quantum gravity, developed during those years in Europe and the United States, no longer speaks of space as a neutral container, but rather as a network of meaningful relationships. It is no coincidence that Rovelli later wrote, many years later, that physics does not tell us *"what the world is like,"* but *"how the world interacts with itself."* A statement that would have made a classical positivist turn up their nose.

The point is not that physics has suddenly become humanistic. Something more interesting has happened: the most advanced models began to work only when they incorporated notions that bear a striking resemblance to what was once called "meaning." In quantum information theory, for example, information is no longer a mere abstract quantity but depends on context, the observer, and the relationship. In Vienna in 2015, during a conference on the foundations of quantum mechanics, an experimental physicist wryly observed that *"information does not exist on its own: it always exists for someone."* A trivial statement for a philosopher, almost scandalous for an old-school physicist.

A similar shift has taken place in neuroscience. For decades, the brain was described as a signal processor: input, processing, output. But starting in the 2000s, especially in laboratories in

Freiburg and San Diego, another idea gained traction: that the brain does not simply process information, but rather produces meaning. The neuroscientist Francisco Varela—who passed away in 2001 but is cited today more than ever—spoke of "sense-making" as the primary function of living systems. This was not a poetic metaphor, but a rigorous attempt to explain why an organism does not merely react, but interprets.

An anecdote shared by an Italian researcher clearly illustrates this shift in perspective. During a study of patients with cortical lesions, she noticed that two subjects with nearly identical damage exhibited radically different behaviors. It wasn't the neural map that made the difference, but their personal history—the way their bodies had learned to inhabit the world. She concluded: *"The brain does not represent the world: it signifies it.*" A conclusion that is difficult to formalize, but impossible to ignore.

It is within this cultural climate that field biology finds its natural place. Not as an invasion of vague concepts into the exact sciences, but as the recognition that living beings cannot be understood without bringing meaning into the picture. Organisms are not simply complex systems; they are structures that maintain meaningful coherence over time. Their information does not serve to describe them from the outside, but to enable them to exist as directed units.

Curiously, this return to meaning does not arise on the margins, but at the heart of the most formalized disciplines. It is as if, after pushing mathematical language to the limits of its power, the hard sciences had stumbled upon a question that equations alone can no longer evade: what truly matters in a living system? Not how big it is, nor how fast it is, but what has been significant for its survival, its form, and its history.

In 2022, at an interdisciplinary conference in Zurich, a theoretical biologist concluded his talk with a phrase that left a lasting impression on many: *"We have learned to calculate everything, except what makes sense.*" It was not a surrender, but an invitation—to recognize that meaning is not a belated addition to science, but rather a dimension that reemerges every time we seriously engage with life.

PART ONE. The Blind Spot of Modern Biology.

From the Thought of Ludwig von Bertalanffy
An organism cannot be explained in terms of its parts, because it is the organization that gives meaning to the parts, not the other way around.

From the thought of Denis Noble:
Genes are essential, but they are not the commanders: life functions as a cooperative system, not as a chain of command.

From the thoughts of Stuart Kauffman:
The living form is not entirely determined by the past: it emerges, stabilizes, and opens up possibilities that did not exist before.

From the thoughts of Henri Bergson:
Life does not simply follow pre-existing laws, but bends them, uses them, and transcends them in its own creative momentum.

The first part of this book opens a critical threshold. Not to demolish what biology has built over the course of more than a century, but to question its foundations where they begin to show signs of strain. The crisis of the classical biological paradigm does not stem from a failure, but rather from a success: the extraordinary effectiveness of biochemical and genetic explanations in describing the local mechanisms of life. It is precisely this effectiveness, however, that makes it more evident what remains outside the picture.

The two chapters that make up this part address a question that is simple only in appearance: what, really, is a living organism?

In the first chapter, the limits of biochemical explanation become apparent when one attempts to move from describing reactions to understanding the unity of the organism. The molecules function, the processes are clear, but the living appears to be more than the sum of its parts: a stable, purposeful whole, capable of sustaining itself and transforming without an external design to guide it. The question of purpose—without resorting to a designer—remains an unresolved issue for classical causality.

The second chapter shifts the focus to Darwinism and natural selection. Here, the crisis does not concern adaptation—which the theory explains with great elegance—but rather form. The emergence, persistence, and coherence of biological structures seem to obey an organizational principle that cannot be reduced either to random variation or to the gene alone. Form does not appear as a secondary effect of function, but as a silent priority that makes every function possible.

Within the book's overall structure, this first part plays a decisive role: it creates the conceptual space necessary for what follows. By showing where the classical paradigm falls short, it prepares the ground for a different conception of life, which will be explored in subsequent parts through the concepts of coherence, field, information, and dynamic equilibrium.

Without this initial crisis, the subsequent proposals would appear arbitrary; with it, becomes a possible answer to questions that have remained unresolved for too long.

This is therefore not a section "against" traditional biology, but one that goes beyond its current boundaries. It is the moment when the reader is invited to set aside the usual explanations—not to abandon them, but to observe them from a critical distance. It is here that the book truly begins to shift its perspective.

In conclusion, this first part invites the reader to recognize that the crisis is not a weakness of science, but a sign that a conceptual transformation is necessary. It is the point at which the question of life once again becomes radical.

I. The Limits of Biochemical Explanation.

This chapter begins with an observation that is as simple as it is unsettling: biochemistry works magnificently, and precisely for this reason, it is now revealing its limitations. No one questions the descriptive power with which molecular biology has elucidated the intimate mechanisms of life—metabolic pathways, enzymatic regulation, gene expression. And yet, when one attempts to move from the "how" to the "what," from local functioning to the organism's overall reality, a discrepancy emerges that can no longer be ignored.

The central argument of this chapter is not a critique of biochemistry, but of its use as a comprehensive explanation of life. Chemical reactions do occur, of course, but they do not, on their own, explain why the organism presents itself as a coherent unit, stable over time and functionally oriented. The cell is not merely a container of reactions: it is a self-sustaining system that selects what matters and acts as if it had a purpose, even without an external plan to guide it.

This is where the classical paradigm begins to show a conceptual flaw. Linear causality—one reaction leading to another—is extremely effective at the local level, but it becomes insufficient when it comes to explaining the unity of the organism. How does the whole emerge from the parts? How do millions of distributed processes give rise to coordinated, purposeful, and surprisingly robust behavior? Biochemistry describes the building blocks, but struggles to account for the building itself.

Even more delicate is the question of purpose. Organisms seem to act "with a purpose in mind": to maintain internal balance, repair damage, or reproduce. Yet modern biology has expelled all references to intrinsic purposes, replacing them with the language of adaptation and selection. The result is an unresolved tension: purpose is everywhere in biological phenomena, but nowhere to be found in theoretical explanations. This chapter openly addresses this paradox, without invoking hidden designers, but by acknowledging that classical causality may not be sufficient.

Within the overall context of the book, Chapter I serves an essential preparatory function. It serves to suspend the evidentia of the dominant paradigm, showing that what appears obvious—life as the sum of chemical processes—is in reality a powerful but incomplete simplification. Only by bringing this limitation into focus does it become possible, in subsequent chapters, to introduce the concepts of coherence, field, and embodied information—not as arbitrary alternatives, but as responses to a real theoretical gap.

This chapter does not ask the reader to abandon scientific explanations, but to recognize that explaining how something happens is not always the same as explaining what it is. It is from this often-overlooked distinction that the entire journey of field biology begins.

Circular causality in living systems: life that causes itself.

If there is one area where biochemical explanations most clearly reveal their limitations, it is precisely in the way they conceive of causality. Classical biochemistry works well when applied to orderly chains of events: a molecule activates an enzyme, the enzyme produces an effect, and the effect triggers another reaction. It is a linear, powerful, and elegant logic. But when applied to the organism as a whole, something doesn't add up. Living systems do not proceed like a row of dominoes; rather, they resemble a system in which what happens alters the very conditions under which it occurs.

This is where the notion of circular causality comes into play—one of the most delicate and least understood concepts in contemporary biology. Circular does not mean vague, much less mystical. It means that, in living systems, causes do not flow in a single direction: the organism's global level influences local processes just as much as local processes contribute to shaping the global level. The organism is not the end result of its reactions: it is the active condition that makes them possible, selects them, and stabilizes them.

This point was already crystal clear to Immanuel Kant when, at the end of the 18th century, he spoke of organisms as beings in which *"the parts exist for the whole and the whole for the parts*." Kant had no access to biochemistry, but he had intuited a problem that has resurfaced with great force today: in living systems, causality is never solely bottom-up. Every cell acts within a context that it did not create on its own, but which it continually helps to maintain.

Circular causality is often confused with simple feedback, a concept borrowed from engineering and cybernetics. The thermostat that regulates a room's temperature is a classic example: it detects a deviation and corrects it. But the organism does something more radical. It does not merely correct a deviation from a preset value; it continually redefines what counts as equilibrium. The set-point is not fixed once and for all: it emerges from the system's own history.

In the 1970s, Humberto Maturana and Francisco Varela introduced the concept of *autopoiesis* precisely to describe this dynamic. A living system—they wrote—is a network of processes that produces the components which, in turn, maintain the network. There is no privileged starting point in the causal chain, because the chain is a circuit. It is an idea that, even today, poses a challenge to any purely linear explanation: who causes what, when everything contributes to everything?

A biologist observing a cell culture knows full well that the behavior of individual cells changes depending on the overall state of the tissue. A doctor monitoring the progression of a chronic disease often observes how the body modifies its response over time, not only reacting to a cause but reorganizing the way it functions. In both cases, causality is not an arrow but a dynamic, historical, context-sensitive circuit.

This idea has profound implications even beyond biology. The philosopher Maurice Merleau-Ponty, reflecting on the living body, spoke of a "logic of mutual implication," in which perception, action, and bodily structure co-determine one another. It is no coincidence that phenomenology and theoretical biology converged on this point in the twentieth century: life eludes any explanation that rigidly separates cause and effect.

Within the framework of Chapter I, circular causality is not yet the solution, but it is already a decisive crack in the classical paradigm. It shows why life cannot be conceived as a sum of independent reactions: because every reaction occurs within an organization that it itself helps to maintain. Biochemistry describes the individual steps; circular causality invites us to look at the dance as a whole.

Words that mean more than they say.

Every science is also an exercise in language. We never describe nature in a neutral way: we recount it through words that, like lenses, bring certain aspects into focus while blurring others. Biochemistry, perhaps more than any other biological discipline, has developed a precise, operational lexicon that appears free of ambiguity. Yet, if we pause to truly listen to this language, a curious tension emerges: the words used to describe molecular reactions seem to allude to something that goes beyond the reactions themselves.

Take common terms such as regulation, control, and signal. No serious biochemist uses them in a metaphorical or intentional sense. Yet these words carry with them a semantic baggage that is not purely chemical. To speak of regulation is to evoke the idea of an order that is maintained; to speak of control suggests a direction, a priority; to speak of a signal implies someone—or something—that receives, interprets, and responds. In other words, the language of biochemistry implicitly introduces the unity of the organism, even when the theoretical paradigm would prefer to limit it to nothing more than the sum of local processes.

Claude Bernard, in mid-nineteenth-century Paris, was perfectly aware of this. When he introduced the concept of *the "milieu intérieur,"* he was not simply adding a technical term: he was recognizing that the living organism actively maintains internal conditions compatible with life. Not an automatic consequence of chemical reactions, but a process of maintenance. Walter Cannon, a few decades later, called this process *homeostasis*, a word that literally means "to remain the same," but which actually describes a dynamic stability that is continually renegotiated. It is hard not to notice how much this idea is already imbued with a unitary view of life.

The same is true when we talk about signals. A signaling molecule is not simply present or absent, like a drop of reagent in a test tube. It is effective only if it is recognized as relevant. In 1956, at the Rockefeller Institute in New York, biologist George Palade demonstrated that cells were not chaotic bags of molecules, but highly organized systems capable of internal trafficking and selective communication. From that moment on, the term "cell signaling" became a permanent fixture in the biological lexicon. But to "signal," even in everyday language, means to indicate something to someone. Biochemistry can describe receptors and intracellular cascades, but the very fact that a molecule counts as a signal depends on the organism's overall state.

Perhaps the most revealing term is "control." We speak of enzymatic control, genetic control, and metabolic control. No one imagines a "conductor" hidden inside the cell, and yet the image remains surprisingly apt. As the evolutionary biologist Conrad Waddington observed during a lecture in Edinburgh in 1968, *"we describe molecular processes as if the organism knew where to go, even when we deny that it knows anything.*" Control, in biology, does not function like a mechanical lever: it is distributed, contextual, and dependent on the system's history. And yet, the word remains, because no other can account for that functional coherence that spans the levels of living organization.

This linguistic tension is not a flaw: it is a symptom. It indicates that biochemistry, despite its localized successes, continually operates on the edge of what it can explain using purely mechanistic categories. Language reveals what the theoretical model struggles to incorporate: the organism as an oriented unit, capable of sustaining itself, responding, and adapting meaningfully to circumstances.

The philosopher of science Evelyn Fox Keller, reflecting on the language of the life sciences, noted how certain terms function as "conceptual bridges": they allow us to work effectively with , while pointing to a broader understanding that remains to be clarified. It is as if biochemistry, in describing individual notes, were compelled to use words that hint at the existence of a melody.

In this sense, the language of biochemistry is not merely a descriptive tool, but a threshold. It reveals that life is not simply what happens when certain molecules react, but what holds those reactions together in a stable, recognizable, and purposeful form. Words know this, even when theory pretends otherwise.

Identity in Pieces: Why Life Cannot Be Taken Apart.

There is a simple, almost trivial gesture that illuminates a profound difference between machines and organisms: disassembly. We can take apart a bicycle in our garage, lay the pieces on the floor, and even then, no one doubts that it remains the same bicycle. It is temporarily unusable, of course, but its identity is not in question. On the contrary, if we take an organism apart—even if we merely separate its vital parts—what remains is no longer that organism. It is not an "incomplete" version: it is something else.

This difference is neither metaphorical nor sentimental. It is ontological. It concerns the way a machine is, as opposed to the way a living being is.

A machine is defined by the functional arrangement of its parts. Its principle of identity is external: it resides in the design, the blueprint, the blueprint that precedes and outlives the concrete object. A diesel engine built in Turin in 1984 is ontologically identical to one assembled today in Pune, provided it meets the same specifications. Dismantling it does not affect its essence: the machine is, after all, a possibility that can always be reconstructed.

An organism, on the other hand, has no external design that defines it, nor any schema separable from its concrete existence. Its identity coincides with its very process of life. As Aristotle already observed in *De Anima*, the living being is not a body that has a form, but a body that is its form in act. Removing its parts means interrupting that process, not suspending it.

Immanuel Kant, in his *Critique of Judgment* (1790), captured this point with a clarity that remains striking to this day: an organism is a being in which "the parts are mutually cause and effect of their form." In a machine, the parts cooperate because someone has arranged them that way. In an organism, the parts exist because the whole is already in act. It is a subtle but decisive difference.

An anecdote often recounted in developmental biology departments dates back to the 1930s, in Hans Spemann's laboratory in Freiburg. When his students separated very early embryonic cells from a salamander, they discovered that each cell could reorganize itself and give rise to a complete organism. But as soon as development proceeded, that possibility vanished: the separation no longer produced two organisms, but rather two fragments destined to perish. They were not "pieces" of a pre-existing whole, like gears on a table; they were moments in a process that exists only as long as it remains intact.

A machine tolerates interruption because its unity is imposed from the outside. The organism does not tolerate it because its unity is immanent. Gilbert Simondon, a French philosopher often overlooked in biology textbooks, explained this in almost engineering terms: the machine is a technical individual that remains so even when inactive; the organism is an individual that exists only in its functioning. To stop it is to dissolve it.

This also clarifies a widespread misunderstanding in the technological metaphors of contemporary biology. Saying that DNA is a "program" or that the cell is a "factory" may be useful in a specific context, but it becomes misleading on an ontological level. A program can be copied, suspended, or restarted. An organism cannot. Not because it is more fragile, but because it is inseparable from its own time. Its identity is historical; it cannot be replicated moment by moment.

Norbert Wiener, the father of cybernetics, admitted this with some unease in a lecture at MIT in 1951: "*A machine can be turned off and on again; an organism cannot, because what we call life is its very continuous becoming.*" It is a statement that sounds almost obvious, yet it marks the exact point where the mechanistic paradigm reveals its inadequacy.

In short, a machine is something that remains what it is even when it is not functioning; an organism is what it is only while it is functioning, growing, and sustaining itself. Dismantling a machine is a reversible operation. Dismantling an organism is an ontologically irreversible event. Not because any parts are missing, but because what held the parts together has ceased to exist: the living unity as a process.

And it is precisely this unity—which cannot be localized anywhere, yet is real—that a biochemical explanation, on its own, fails to capture.

The Local Success of Biochemistry.

If there is one discipline that, in the twentieth century, has lived up to the promises of modern science, it is biochemistry. It has explained, predicted, and manipulated. It has transformed medicine, agriculture, and the food industry. It has given names, formulas, and pathways to what had remained obscure for centuries. It would be intellectually dishonest not to acknowledge its success. The problem is not what biochemistry explains, but what remains outside the scope of its explanations.

When, in 1926, James Sumner crystallized urease, demonstrating that an enzyme is a protein, he put an end to a dispute that had seemed almost metaphysical. From that moment on, life appeared increasingly as a matter of catalyzed reactions, molecular affinities, and measurable kinetics. From there on, the story is well known: the discovery of the structure of DNA at Cambridge in 1953, the genetic code deciphered in the 1960s, metabolic pathways drawn like subway maps in university textbooks.

In this sense, biochemistry works flawlessly. It works when it isolates a process, reproduces it in vitro, and measures the conditions under which it occurs. It works when a drug blocks a receptor, when an inhibitor slows down an enzymatic cascade, when a mutation is linked to a defective protein. It is an extraordinarily effective science at the local level: the active site of an enzyme, the binding pocket of a ligand, the sequence of a polypeptide chain.

A researcher at EMBL in Heidelberg remarked during an informal seminar in the early 2000s that " *, everything adds up* *in the lab—until you put the molecule back into the cell.*" It wasn't a jab at biochemistry, but an acknowledgment of its limitations. In a test tube, the system is clean, controllable, and elegantly reducible. In a living organism, the same reaction is immersed in a context that modifies it, diverts it, and sometimes subverts it.

The crucial point is that biochemistry describes interactions, not organizations. It can explain how one molecule recognizes another, but not why that encounter occurs there, at that moment, in that sequence, contributing to the stability of the entire organism. The sum of the correct reactions does not, by itself, guarantee the presence of life. As François Jacob, the 1965 Nobel laureate, observed, "*Biology is not a physics of the living, but a story.*" And a story cannot be reconstructed by listing the letters of the alphabet.

An emblematic example comes from molecular biology in the early 1970s. At the Salk Institute in La Jolla, several research groups succeeded in reconstructing fundamental processes such as transcription and translation in vitro. Everything worked. But when they attempted to apply that explanatory framework to the entire organism, a subtle yet persistent discrepancy emerged: the same molecules, in the same quantities, did not produce the same result. Something depended on the overall context, on the dynamic organization, on relationships that could not be reduced to a linear causal chain.

Biochemistry excels at answering the question "How does a reaction occur?", but struggles when the question becomes "Why is this reaction relevant to the organism's life?". It is a difference of scale and logic, not of precision. The reductionist paradigm is not false: it is locally true. The problem arises when its success is improperly extended to what is not local.

In the 1980s, the English biologist Conrad Waddington—a figure less frequently cited today than he deserves to be—observed that biochemistry had explained the "building blocks" of life, but not its "architecture." And he added, with a touch of irony, that no one would claim to understand a Gothic cathedral by analyzing only the chemical composition of its stones. The comparison does not diminish the chemistry of the stones; rather, it highlights the limits of what it can explain.

The success of biochemistry, therefore, is real and measurable. But it is a success that illuminates certain aspects of the phenomenon of life while leaving other, crucial aspects in the shadows. As long as we remain at the level of reactions, everything is clear. When we ask ourselves how those reactions hold together, stabilize, and coordinate over time, the language of biochemistry suddenly falls silent. Not because it is wrong, but because it was not designed to answer that kind of question.

And it is precisely in this gray area—between local effectiveness and global inadequacy—that the need for a different perspective on life begins to emerge.

From molecules to networks: when life changed scale.

For much of the twentieth century, biochemistry represented the beating heart of the life sciences. Not only because of the results it yielded, but because of the very idea of knowledge it embodied: understanding an organism meant breaking it down into its elementary components, studying them one by one, and then—at least in principle—reconstructing the whole. An ambitious program, consistent with the spirit of scientific modernity, and in many ways extraordinarily productive.

In the laboratories of Harvard, Cambridge, or the Pasteur Institute in Paris, between the 1950s and 1970s, this approach seemed to have no rivals. Every new metabolic pathway mapped, every receptor identified, every enzyme characterized reinforced the idea that life was, at its core, a matter of well-organized chemistry. As an elderly researcher at the Karolinska Institutet ironically remarked, *"If we don't understand it, it's only because we haven't found the right molecule yet."*

Yet, just as this approach was at the height of its success, the first conceptual cracks began to appear. They did not stem from an ideological rebellion, but from day-to-day laboratory experience. By the late 1980s, with the advent of genomics and proteomics, it became clear that knowing all the components did not equate to understanding the system's behavior. Genomes were being sequenced, lists of proteins were growing longer, but the gap between inventory and understanding was not narrowing.

An episode often recalled—almost with embarrassment—dates back to 1995, during a conference in Santa Fe, a symbolic location for the complex sciences. Faced with extremely elegant mathematical models and an impressive amount of molecular data, someone pointed out that while it was possible to describe a cell as a network of thousands of interactions, it was not possible to reliably predict how it would react to even the slightest disturbance. The joke that circulated in the hallways was: *"We know everything, except what it does*."

It is in this climate that systems biology took shape—not as a rejection of biochemistry, but as its critical extension. The continuities are evident: the same data, the same tools, the same emphasis on quantitative rigor. But the shift lies in the perspective. No longer is the individual reaction the fundamental explanatory unit, but rather the set of interactions considered simultaneously, across time and space.

In 2000, in Tokyo, Hiroaki Kitano—a key figure in this transition—explicitly spoke of the need to move from a "biology of parts" to a "biology of emergent behaviors." The term "emergence" had nothing mystical about it: it simply referred to observable properties that cannot be attributed to any single isolated component. This was an empirical observation, even before it was a theoretical one.

Systems biology thus arose from a dissatisfaction with the previous paradigm. It does not deny that chemical reactions are real, necessary, and fundamental. But it shows that their biological significance changes depending on the network in which they are embedded. The same molecule can act as a signal, a nutrient, or a toxin, depending on the global state of the system. Here, linear causality—so effective in classical biochemistry—begins to reveal its limitations.

It is interesting to note that this shift also has broader cultural implications. In the very same years that biology discovered networks, the Internet became a dominant metaphor; economics began discussing complex systems; and ecology abandoned simplistic models to describe dynamic equilibria. This is no coincidence: it is a sign of a shift in scientific sensibility, of a new focus on relationships rather than on objects.

And yet, systems biology remains—deliberately—within the realm of experimental science. It does not introduce new entities, nor does it invoke external principles. It merely states that life cannot be explained by simply changing the level of detail, but requires a shift in perspective. It is a conceptual threshold, not yet a complete new paradigm. But it is a decisive threshold.

In this sense, the transition from classical biochemistry to systems biology represents a moment of suspension: the realization that the language of reactions is no longer sufficient, without it yet being clear what language will replace it. A fertile, often restless transitional phase in which science begins to intuit that the organism is not merely a set of processes, but a dynamic unity that gives meaning to its own mechanisms.

The whole that cannot be broken down.

There comes a moment in the study of life when even the most refined analysis comes to a halt in the face of a seemingly naive question: where is the organism? Not in individual molecules, not in isolated genes, not in tissues observed as separate compartments. And yet, the organism is there; it asserts itself, resists, and is recognized as a coherent unit over time. Modern biology, extraordinarily skilled at describing details, still struggles to explain this "something more" that turns a collection of parts into a living being.

The problem is not new. As early as the beginning of the nineteenth century, Johann Wolfgang von Goethe, observing the metamorphoses of plants in his garden in Weimar, noted that living form is never a simple result, but a process that expresses itself. He did not seek the "true" plant in a single privileged organ, but in the movement that connects them all. A leaf is not just a leaf: it is a temporary variation on a unified theme. Goethe did not speak the language of biochemistry, but he had grasped something that still challenges us today.

Classical reductionism proceeds by decomposition: it understands a function by isolating it. But the unity of the organism does not behave like a function. It is a distributed presence that never coincides with a precise point. Kurt Goldstein, a German neurologist forced into exile in the United States in the 1930s, understood this while working with brain-injured soldiers from World War I. Similar injuries produced different outcomes in different people. It was not the missing piece that explained the deficit, but the reorganization of the entire organism around that loss. Goldstein spoke of the organism as a totality, a concept that mainstream neurology of the time considered embarrassingly vague.

Yet the facts were stubborn. An organism reacts as a whole even when one part is affected. It does not simply sum up local responses: it redefines itself. This is the blind spot of the purely biochemical explanation. The reactions are necessary, but they do not explain why they occur in a coordinated, stable manner oriented toward the preservation of unity.

In the 1950s, working between Brussels and Austin, Ilya Prigogine showed that certain physical systems far from equilibrium can self-organize. But what was striking about living beings was the persistence of identity: not just order, but an order that recognizes and maintains itself. A vortex in a fluid exists as long as certain conditions persist; an organism, on the other hand, actively contributes to producing the conditions of its own existence. Here, physics comes close to the boundary, but does not cross it entirely.

An anecdote often recounted in embryology laboratories concerns Hans Driesch, who in 1891, in Naples, was working on sea urchin eggs. He separated the early embryonic cells, expecting to obtain incomplete fragments. Instead, he obtained entire organisms, albeit smaller ones. Driesch, initially a staunch mechanist, was forced to change his mind: there was no single "part" in charge, but rather a global capacity for organization that could not be pinpointed. He spoke of *"entelechy,"* a term now out of fashion but coined to address a real problem.

The crucial point is that the unity of the organism does not emerge after the parts are assembled, as happens with an artifact. It is present from the very beginning as an operative principle, as a context that gives meaning to the parts as they differentiate. In this sense, the whole is not the final result: it is the initial condition that makes every differentiation possible.

The philosopher Maurice Blondel wrote in 1893 that *"life is an action that can only be understood from within."* Today we would say that the unity of the living is a unity of meaning, not merely of structure. A liver is not "a liver in general," but that liver, in that body, within that biological history. Remove it from there, and what remains is competent but mute matter.

The problem of the organism's unity, therefore, is not a temporary flaw in biological theory, awaiting better data. It is a conceptual challenge. It calls for us to conceive of the living not as a tightly fitted mosaic, but as a dynamic form that holds together difference and identity, transformation and permanence. A structure that is not found anywhere in particular, but manifests itself everywhere.

The face we recognize without knowing why.

There is an everyday experience we are all familiar with that, when observed carefully, reveals a great deal about the nature of the living organism. We recognize a face from a distance—perhaps in a blurry photograph or after years of absence—without being able to pinpoint precisely which features make it unmistakable. It is not the nose, nor the eyes taken individually, nor their simple geometric combination. It is something that lies "between" the parts—or rather, that runs through them all. A perceived unity, not a measured one.

This aesthetic experience—for it is indeed aesthetic, in the original sense of *"aisthesis,"* or perception—is surprisingly close to what happens when we encounter a living organism. Here, too, the whole manifests itself before analysis. No biologist, when observing an animal or a plant, needs to list its molecules to recognize its identity. The organism presents itself as a unity, even before it is explained.

In 1896, in Paris, Henri Bergson attended a theatrical performance and noted in his notebooks how an actor's gesture was recognizable even when performed by another body. It was not the sequence of muscle movements that mattered, but the style of the movement. Bergson would use this example to speak of life as élan, but the observation remains valid: style precedes decomposition. Similarly, an organism has a "biological style," an overarching signature that does not coincide with any isolated element.

Consider music. A Bach fugue remains recognizable even when transposed to another key or played on different instruments. The individual notes matter, of course, but they do not explain the unity of the work. Musical unity is a relational structure, an internal coherence that is maintained through variations. Similarly, an organism undergoes continuous changes—cell turnover, physiological adaptations, cycles—without losing its identity. The parts change; the whole remains.

An anecdote often recounted in art history circles concerns Ernst Gombrich, who, in the 1960s in London, showed his students deliberately imperfect copies of famous paintings. Many recognized the artist even in the presence of obvious errors. "Style," Gombrich would say, "is not a detail; it is a totality that filters every detail." This statement could easily be applied to biology: the living form filters and organizes every local process.

Biochemistry, like the strictest formalist criticism, tends to ask what something is made of. But aesthetic unity—like biological unity—answers a different question: what makes this what it is? When we look at a painting by Morandi, for example, we do not reduce it to a list of the objects depicted. Bottles and boxes can be found elsewhere as well. What we recognize is a compositional tension, a silent balance that cannot be translated into independent parameters.

In the 1980s, studies in Gestalt psychology—conducted in Berlin and New York—showed that the perception of the human face resists fragmentation: even when certain features are altered or hidden, the face remains identifiable as long as the overall configuration is preserved. This is a valuable lesson for biology: unity is not a side effect of the parts, but a primary perceptual and functional condition.

The organism, like a successful work of art, is not merely complex. It is coherent. And coherence can never be pinpointed. It does not reside in a gene, just as it does not reside in a brushstroke. It is a property of the field of relationships that holds the whole together. Understanding this means recognizing that life cannot be explained solely by an inventory of its components, just as art cannot be understood solely through an analysis of its materials.

In this sense, biological and aesthetic unity share the same profound logic: both exist as forms that are recognizable before they are describable, as presences that precede measurement. They do not deny the importance of the parts, but subordinate them to an order that does not arise from the sum, but rather guides it.

When Cells "Know" What to Do: The Challenge of Purpose Without a Designer.

There is a question that has accompanied biology since its modern origins and which, despite centuries of discoveries, continues to unsettle laboratories and university classrooms: why do organisms seem to act "with a purpose in mind," even though they were not designed by anyone? The heart of the problem does not lie in how the mechanisms of life function—biochemistry has worked wonders in this regard—but in why those mechanisms cooperate in such a coherent, stable, and purposeful way.

When we observe a wound healing, an embryo developing according to a surprisingly reliable sequence, or an immune system that "recognizes" what needs to be eliminated, we are led to use teleological language: to heal, to develop, to defend oneself. Yet, ever since Darwin, science has done everything in its power to expel finality from its official vocabulary. Speaking of purposes seemed like a disguised return to the natural theology of William Paley, famous for his example of the watch found in a meadow: if there is a watch, there must be a watchmaker. But modern biology wants no watchmakers.

The result is a tension that has never been fully resolved. On the one hand, organisms behave as if they had a purpose; on the other, the only causality accepted is efficient causality: chains of local, blind, unintentional events. The language of textbooks therefore seeks compromise solutions: "teleonomy," "selected function," "genetic program" (n informatics metaphor that

Jacques Monod used with caution as early as the 1960s). But the unease remains.

An anecdote recounted in the 1990s at the embryology laboratory at the University of Cambridge is emblematic. During an informal discussion, a young researcher, observing the regeneration of a salamander embryo from which a part had been removed, blurted out: *"It seems to know what's missing."* The remark was met with wry smiles. Yet no one could offer an explanation that did not implicitly resort to that forbidden idea: a global form guiding local processes.

Classical causality works beautifully when it comes to explaining how a reaction occurs or why a protein binds to a receptor. But when the subject is the organism as a whole, the explanation breaks down. How can a sequence of local causes produce behavior oriented toward maintaining form, integrity, and overall functionality? To say that it is "merely" the result of natural selection shifts the problem in time, but does not eliminate it. Selection explains why certain configurations persist, not how they operate, here and now, as purposeful units.

Immanuel Kant, in *his *Critique of* Judgment* (1790), had already grasped this point with a clarity that still surprises us today. Organisms, he wrote, are entities in which the parts exist for the whole and the whole for the parts. They are not like a machine, assembled from the outside, but like a reality that produces itself. Kant concluded that finality in biology is inevitable as a principle of judgment, even if it cannot be demonstrated as a real cause. Two centuries later, the problem is still there, merely cloaked in more technical language.

Today we avoid words like "entelechy," but the phenomenon has not disappeared. In systems biology laboratories, in Boston as well as in Zurich, researchers study genetic and metabolic networks that exhibit robustness: the ability to maintain a function even when individual components are disrupted. It is an elegant, mathemat l language, but beneath the surface, the same question resurfaces: robustness with respect to what? To a form, a function, or an expected outcome.

The paradox of purpose without a designer is this: life seems oriented without being intentional, directed without a director. Classical causality—linear and local—lacks the conceptual tools to account for this intrinsic directionality. It can describe the path, but not the reason for the path. It can list the stages, but cannot explain why, despite noise, errors, and perturbations, the organism always returns—as long as it is alive—to a certain functional coherence.

It is no surprise, then, that in everyday practice, even the most rigorous scientists slip into teleological language. They speak of "errors to be corrected," "appropriate signals," and "adequate responses." It is operational language, to be sure, but it betrays something deeper: the difficulty of conceiving of life without attributing to it—at least implicitly—an internal meaning.

Perhaps the enigma lies not in the fact that biology has not yet managed to explain purpose, but in the assumption that every explanation must be reduced to blind and local causes. As the philosopher of science Francisco Varela observed in a lecture in Paris in 1996, "the question is not to reintroduce a designer, but to understand how normativity emerges without a legislator." A statement noted by many, but rarely taken to its full logical conclusion.

Purpose without a designer thus remains a frontier zone: too empirical for traditional metaphysics, too rich in meaning for a biochemistry reduced to a set of cogs. It is precisely within this tension that space opens up for a new question about life—no longer just what it is made of, but what kind of order it is.

The purpose without a purpose: re-reading Aristotle in the light of contemporary life.

When Aristotle's name is mentioned in a molecular biology department, a polite silence often falls, followed by a quick change of subject. Yet, without our realizing it, many of the conceptual difficulties in contemporary biology still revolve around a notion that the philosopher from Stagira had placed at the center of his reflection: the final cause, the *"télos."*

Aristotle did not conceive of finality as a project imposed from the outside, nor as the execution of a conscious plan. In **De Partibus* Animalium*, he insists on a point that sounds surprisingly modern today: in living organisms, structures exist in view of a function, and this function is internal to the very process of development. There is no hidden architect, but rather an intrinsic orientation of form toward its own realization.

Modern biology has reacted to this idea with a long process of repression. From the seventeenth century onward, with Descartes and the birth of mechanistic science, the final cause was progressively expelled from scientific discourse, considered a metaphysical shortcut. To explain means to identify efficient causes, chains of measurable events. Everything else—purposes, ends, tendencies—is dismissed as poetic language or, worse, as theological residue.

Yet, on closer inspection, the problem Aristotle sought to address has never disappeared. How can we explain that a mammalian embryo, starting from a single cell, undergoes a highly regulated sequence of transformations, almost always—barring accidents—resulting in a coherent functional form? Saying that "it happens because it is selected" is not enough to clarify why, during development, each step seems oriented toward a future configuration that does not yet exist.

An anecdote recounted by the historian of science Pietro Corsi concerns a lecture given in Paris in the 1970s by a well-known evolutionary biologist. When a student asked how to avoid teleological language when discussing development, the professor replied with a smile: *"We avoid it in article titles, not in the hallways."* A quip that captures the situation well: teleology is officially banned, but it continues to operate unofficially.

Rereading Aristotle today does not mean adopting his cosmology or descriptive biology—which have been outdated for centuries—but rather recognizing the relevance of his central insight: in living beings, the present seems organized in function of a future form. The final cause, viewed in this light, is not a mysterious force, but a way of describing a reversed temporal order, in which the meaning of a process cannot be reduced to the sum of its intermediate steps.

In the twentieth century, some thinkers attempted to reopen the dialogue with Aristotle without falling into naive vitalism. The French philosopher Raymond Ruyer, for example, spoke in the 1950s of "formal activity" to describe those biological processes that cannot be explained solely as local mechanical effects. His ideas remained marginal, but they anticipated a sense of unease that resurfaces today in discussions of morphogenesis and self-organization.

The crucial point is that classical causality always looks backward, seeking causes in the immediate past. The Aristotelian final cause, on the other hand, looks forward: not as a conscious prediction, but as an attraction toward a stable configuration. In modern terms, we might say that Aristotle was attempting to name what we today describe, using different terminology, as "dynamics toward organized states."

An often-overlooked detail is that Aristotle was a tireless observer. On the island of Lesbos, where he spent long periods studying marine animals, he noted variations, anomalies, and borderline cases. He was not seeking abstract perfection, but regularity within imperfection. For him, the final cause was not a metaphysical guarantee, but an interpretive hypothesis to make the overall behavior of living beings intelligible.

Viewed from this perspective, the distance between Aristotle and field biology—which interprets the organism as a stabilized informational structure—narrows. In both cases, what matters is not the individual component, but the overall form that guides the processes. The words change, but not the underlying problem. And perhaps, as often happens in the history of science, an ancient idea resurfaces not because it was "true" in the modern sense, but because it had identified a question we have not yet stopped asking ourselves.

The final cause, freed from the idea of an external designer, could then be reinterpreted not as the ultimate explanation, but as a sign of a limit: the point at which a purely mechanical description ceases to be sufficient. Aristotle did not have the tools to go beyond that. We do, perhaps. But only if we are willing to acknowledge that life poses questions that cannot easily be reduced to reactions and causal chains.

Diaries, letters, silences: the purpose that recurs in scientists' confessions.

There is a moment in scientific practice that rarely appears in *peer-reviewed* articles but surfaces with surprising regularity in private letters, laboratory journals, and interviews given at the end of a career. It is the moment of hesitation—the moment when the researcher senses that the purely mechanical explanation, however technically flawless, fails to say everything that seems necessary to say.

François Jacob, winner of the 1965 Nobel Prize in Medicine, admitted this with disarming candor in some autobiographical pages published in 1970 (**La logique du vivant**): *"We use language that attributes purposes and intentions— —even when we know we shouldn't.*" Jacob was fully aligned with the molecular paradigm, yet he recognized that terms such as "function," "program," and "control" were not merely innocent metaphors. They were conceptual crutches, necessary for navigating a territory where linear causality showed obvious cracks.

This unease does not stem from metaphysical nostalgia, but from the concrete experience of scientific work. It can be found, for example, in the laboratory notebooks of Conrad Hal Waddington, a British embryologist, who in the 1940s jotted down reflections in the margins of his experiments on *Drosophila* that were unusual for a geneticist: *"Development seems to remember where it is supposed to go*." Waddington coined the term *"epigenetic* landscape" precisely to avoid teleological language, but the metaphor of the "channel" and the "valley" betrayed an unresolved tension: how can one speak of direction without speaking of purpose?

Even in the great temples of experimental biology, this awkwardness was palpable. At the *Cold Spring Harbor* Laboratory in New York State in the 1950s, some young researchers recounted—in interviews collected decades later by science historians—that during informal evening discussions they spoke freely of cellular "goals," only to then carefully purge that vocabulary from manuscripts intended for journals. It was a sort of dual register: a practical one, indispensable for thinking; an official one, necessary for publication.

The Swiss biologist Adolf Portmann, less well-known to the general public but widely read in the 1950s, openly addressed this discrepancy. In his personal notebooks, now preserved in Basel, he wrote that the organism "reveals itself," as if form had an expressive value as well as a functional one. He was not speaking of purpose in the classical sense, but he acknowledged that reducing form to a mere adaptive byproduct left a sense of incompleteness. His views earned him a certain degree of academic marginalization, but also the attention of philosophers and art historians.

The interesting point is that the unease does not stem from ignorance of the mechanisms, but rather from their narrative inadequacy. Scientists know how many processes occur; they struggle to explain why these processes converge so reliably toward stable functional configurations. In a 1996 interview with the BBC, developmental biologist Anne McLaren, a leading figure in mammalian embryo research, confessed that the term "cell fate" struck her as "dangerous but inevitable." Dangerous because it evoked teleology, inevitable because no other term worked quite as well.

Admissions of this kind are not uncommon, but they remain confined to the margins of public discourse. It is as if contemporary biology were subject to a form of epistemological self-censorship: it acknowledges certain problems but avoids articulating them fully for fear of straying beyond the boundaries of "hard" science. Yet it is precisely these hesitations, these emotional footnotes, that point to the point of greatest tension within the biochemical paradigm.

In the context of field biology, this unease takes on new meaning. If the organism is a stabilized informational structure, then the recurrence of purpose-oriented language is not a subjective weakness of the scientist, but a symptom of a level of organization that demands different categories. Letters, diaries, and interviews thus become valuable documents: not deviations from the method, but traces of a real, still-open problem.

Perhaps, as the physicist Max Delbrück wrote in a private letter in 1972 to a colleague in Pasadena, *"we have won many battles against mystery, but not the war against meaning.*" A phrase we will never find in a textbook, but one that says a great deal about what still eludes us when we try to explain life without resorting, at least implicitly, to some form of purpose.

When systems do more than we programmed them to.

In recent years, a curious reversal of perspective has crept into research laboratories: it is no longer just biology that inspires engineering, but the engineering of autonomous systems that is forcing biology to reevaluate its own certainties. In particular, artificial intelligence—the real kind, made up of code, robots, and simulation environments—has brought back into focus an issue that classical biology had sought to set aside: the emergence of goal-oriented behaviors without the goal having been explicitly programmed.

An anecdote often told in the halls of MIT concerns an autonomous robotics project from the early 2000s. A group led by Rodney Brooks, a pioneer in behavioral robotics, had designed small mobile robots with very simple rules, lacking any internal representations of the world. During testing in a warehouse in Cambridge, Massachusetts, some of these robots spontaneously began to "defend" certain areas rich in energy resources, driving the others away. None of the instructions mentioned defense, competition, or territory. Yet the behavior was there, stable and repeatable.

The engineers called it *"emergent behavior."* But among the biologists present at the demonstrations, a more unsettling question was circulating: isn't this exactly what we observe in living organisms? We're referring to *"reinforcement learning,"* a term that attributes goals, intentions, or objectives to the agent, even though in reality the behavior emerges solely from update rules and reward signals.

In this field, the issue has become even more apparent. When AlphaGo defeated Lee Sedol in Seoul in 2016, many commentators spoke of the machine's "creativity." The programmers, more cautious, emphasized that the system was simply maximizing an objective function. But in subsequent interviews, Demis Hassabis admitted that some strategies were neither anticipated nor understood a priori: they emerged from the interaction between the algorithm, the environment, and the history of the games. An operational purpose without a conscious representation of the goal.

The parallel with biology is hard to ignore. In organisms, too, purpose does not appear as a plan written down somewhere, but as a stable tendency of the system to maintain certain forms, certain functions, and certain relationships. AI shows that it is possible to achieve functional orientation without resorting to a designer who controls every step. But it also reveals something more subtle: that this purpose cannot be reduced to the sum of local rules.

In a lecture given in Santa Fe in 2011 at the Santa Fe Institute—a symbolic hub for the study of complex systems—physicist Chris Langton observed that some cellular automata "seem to know what to avoid." It was a joke, but not entirely. Formal systems, governed by elementary rules, developed dynamics of preservation and adaptation that resembled—at least by structural analogy—the behavior of living systems.

Here, the connection with field biology becomes particularly fruitful. If, in organisms, purpose is an emergent property of the field of relationships that constitutes them, then AI provides a conceptual laboratory in which to observe how goal-orientation can arise from global configurations, rather than from specific instructions. The machine does not "want" anything, and yet it behaves as if it were striving toward something. Just like a cell healing a wound or an embryo correcting a disrupted development.

It is no coincidence that some researchers, such as the British mathematician Michael Levin—known for his studies on morphogenesis—have begun to use conceptual tools borrowed from AI to describe biological processes, speaking of "goal-directedness" at different levels of organization. This is not a matter of attributing intentions, but of recognizing a structure of meaning and operat —a directionality that cannot be localized to a single molecule, nor to a single line of code.

The paradox is evident: while classical biology has long avoided the language of purpose in order to remain "scientific," the engineering of autonomous systems now finds itself forced to grapple with purposeful behaviors that no one explicitly intended to introduce. It is as if two different paths were converging toward the same blind spot of reductionism.

In this sense, artificial intelligence does not offer a model to copy, but a mirror. It shows that purpose can emerge without intention, and that the problem is not to eliminate purpose from scientific discourse, but to understand at what level it belongs. This is a valuable lesson for a biology that seeks to conceive of organisms not as complicated machines, but as stabilized informational structures capable of maintaining a direction over time.

II. When Evolution Plays, but Doesn't Write the Score.

There is an image that helps more than many formulas in understanding the heart of this chapter: that of a blind person listening to an orchestra. He can recognize that the music is harmonious; he can even distinguish the instruments and appreciate the balance among the parts. But he does not see the score, nor the principle that holds the performance together. Natural selection, in the classic account of Darwinism, closely resembles that blind man: it explains why a certain variation is favored, but not why there is a coherent form to select from.

Darwinism has been, and still is, one of the most powerful scientific theories ever formulated. It has shown how the adaptation of organisms to their environment can emerge without resorting to preestablished designs or external teleology. Yet its very success has ended up obscuring a deeper question: where do biological forms come from? Not the individual variations, but the stable architecture that makes variation itself possible.

Natural selection always acts upon something: upon organisms already endowed with a structure, a symmetry, and a robustness that resists disturbances. A heart may vary in size, a limb in length, a beak in curvature, but the presence of the heart, the limb, or the beak is not explained by the selective mechanism as such. Selection filters, refines, and eliminates. It does not orchestrate.

Charles Darwin was perfectly aware of this, even though it is often forgotten. In **On the Origin of* Species*, he writes, with the caution typical of his style: *"We are led to personify the law of variation, as if it were an independent force; but we must remember that variation is not the cause of the production of new species, but merely the raw material upon which selection acts."*

A sentence that, read today, sounds almost like an admission of incompleteness: where does this "raw material" come from? And above all, why does it take on such regular and recurring forms?

This chapter is situated here, in this conceptual space left in the shadows. Not to deny Darwinism, but to put its explanatory scope into perspective. Selection explains local adaptation, not the global stability of forms. It explains why something is advantageous, not why it exists. As the English biologist W. H. Thorpe observed, reflecting on the limits of the adaptationist approach: *"The problem of the origin of species concerns not only adaptation, but also the origin of complex and highly specific structures that make adaptation itself possible."*

This is where the problem of form as an autonomous organizing principle emerges. Biological forms display a surprising ability to maintain, regenerate, and correct themselves. A disturbed embryo tends to "rediscover" its own organization; an injured organism seeks to reestablish a recognizable configuration. All of this occurs prior to and independent of selective pressure. Form is not a late result of evolution, but a condition of its possibility.

Henri Bergson, at the beginning of the 20th century, had intuited this with philosophical clarity when he wrote in *L'évolution créatrice*: "*Life is not a mechanism; it is a tendency. A force that seeks to overcome the resistance of matter and to create, at the very heart of matter, the instrument of its own evolution.*" Beyond the metaphorical language, the insight is precise: life does not proceed through the simple accumulation of selective advantages, but through formal configurations that guide and channel change.

The purpose of this chapter, within the broader context of the book, is therefore clear: to show that the crisis of the classical biological paradigm concerns not only molecular mechanisms but also the very heart of evolutionary theory. Without a reflection on form, selection remains like a conductor without a score: capable of keeping time, but unable to explain the music.

This step is crucial for paving the way for the subsequent sections of the book. If form cannot be reduced to either genes or selection, then we must look elsewhere for the organizing principle of life: in fields of coherence, in informational structures, in an order that is not imposed from the outside but emerges as a global property of biological systems.

The Form That Endures: The Mystery of Biological Robustness.

There is one aspect of life that strikes anyone who observes it without theoretical bias: its surprising ability to remain true to itself while everything else changes. Organisms develop in imperfect environments, face disturbances, errors, and molecular fluctuations, yet they almost always arrive at a recognizable, stable, functional form. This property, which biologists call robustness, is one of the most glaring blind spots of classical Darwinism, because it lies precisely between two domains that evolutionary theory has struggled to integrate: developmental biology and evolution.

Natural selection can explain why a robust form is favored: an organism that "holds up" better survives more easily. But it does not explain why that robustness emerges in the first place, nor how a form manages to maintain its identity during development, despite genetic variations, biochemical noise, and changing environmental conditions. This is where evolution encounters a paradox: in order to vary, a form must first be stable; yet stability itself cannot be the direct product of random variation.

In the 2000s, systems biologist Hiroaki Kitano, working between Tokyo and Europe, defined robustness as a fundamental property of living systems, one that cannot be reduced to individual genes or isolated circuits. In a now-classic 2004 article, he wrote: *"Robustness is the ability of a system to maintain its functions despite internal and external perturbations*." A seemingly understated definition, yet one that radically reframes the problem: form is not a passive outcome, but an actively maintained configuration.

From a developmental perspective, this means that the organism does not follow a simple sequence of instructions, like a linear program, but rather orients itself toward a global configuration, correcting deviations, compensating for errors, and locally reorganizing whatever has been disrupted. From an evolutionary perspective, it means that selection does not act on amorphous material, but on forms already endowed with intrinsic resilience, which limits the range of possible variations and channels their effects.

A similar insight was already present, in a pre-molecular form, in the work of D'Arcy Wentworth Thompson, who in 1917, in **On Growth and* Form*, observed that many biological structures seemed to obey constraints of order and stability rather than a simple adaptive logic. Thompson did not deny evolution, but suggested that the geometry and dynamics of forms precede natural selection, providing it with the ground on which to operate.

Robustness, in this sense, serves as a bridging concept: it connects individual development to evolutionary history, demonstrating that life is not a sequence of provisional solutions, but a system capable of preserving structural information over time. Biological forms are not fragile equilibria on the verge of collapse, but active configurations that resist, repair themselves, and reorganize. And it is precisely this intelligent resistance to disorder that makes evolution itself possible.

In the context of this chapter, robustness thus becomes a decisive argument against a purely selective view of form. If forms were not robust, selection would have nothing stable upon which to act. The orchestra—to return to the chapter's metaphor—could not even begin to play if every instrument fell apart at the first stroke of the bow.

Vortices, snowflakes, and cells: deceptive similarities in form.

To understand what makes a form biological, it is helpful to take a step back and look at the forms that nature produces even in the absence of life. Crystals, vortices in fluids, and the so-called dissipative structures studied by twentieth-century physics show that order can emerge spontaneously from chaos, without the need for an external design. It is a discovery that has fascinated generations of scientists and that, at first glance, seems to bring the inanimate world closer to the living one. But it is precisely in this convergence that the distance becomes clear.

A snowflake, viewed under a microscope, appears as a tiny masterpiece of symmetry. Wilson Bentley, the Vermont photographer who devoted his life to photographing them in the early 20th century, loved to say that *"no two snowflakes are alike.*" Yet, behind that variety lies an ironclad rule: the crystal's shape is entirely determined by the physical laws and by local conditions of temperature and humidity. The crystal does not "defend" its shape, does not correct it, does not interpret it. If conditions change, the shape changes or dissolves. Stability is passive, not active.

We find a step closer to life in vortices: a spiral of water in a river, a red spot on Jupiter already observed by Cassini in the 17th century, or the convection cells described by Henri Bénard in 1900 in his Parisian experiments. Here, the shape is no longer static: it is a process that is sustained over time thanks to a continuous flow of energy.

The analogy with living organisms is compelling. Life, too, is an open system, far from equilibrium, that sustains itself through continuous exchanges with the environment. But the crucial difference emerges as soon as we look at how form is preserved. A vortex has no memory: if the flow is interrupted, the form disappears without a trace. An organism, on the other hand, endures flows, disturbances, even traumas, and tends to return to a recognizable configuration, as if the form were inscribed at a deeper level than the mere movement of matter.

Here the limit of the physical analogy becomes apparent. Physical forms are epiphenomena of conditions: they result from the laws, but do not act upon them. Biological forms, on the contrary, seem to exert a sort of top-down causality, directing local processes toward a global outcome. During embryonic development, cells that are genetically identical take on different fates not because the physical context dictates it, but because the organism as a whole "knows"—in an operational, not metaphorical sense—where it must go.

This comparison also clarifies a central point: natural selection can act on pre-existing forms, but it is not sufficient to explain why biological forms possess this active, regulatory, and persistent quality. Physical structures show that order is possible without selection; life shows that there is a type of order that resists, repairs itself, and anticipates the future. It is in this subtle but decisive difference that biological form reveals its conceptual autonomy.

The orchestra of life, therefore, is not made up solely of sound physical laws that play together by chance. There is an implicit score that runs through the processes, coordinates them, and makes them coherent over time. And it is precisely this score—which is still difficult to name without straying beyond the bounds of reductionism—that distinguishes a snowflake from a cell, a vortex from an organism.

When Form Demands a Deeper Principle.

At this point, the attentive reader senses a tension that cannot remain unresolved. We have seen that natural selection explains how one form functions better than another, but not why that form exists, nor why it manages to remain stable over time, enduring development, mutations, and environmental stress. The metaphor of the orchestra, evoked in the chapter's title, then returns with force: it is not enough to imagine well-crafted instruments or musicians selected for their skill. Without a principle that holds them together—that synchronizes tempos, intensities, and relationships—what emerges is not music, but noise.

This principle is not a new substance, nor is it a mysterious entity to be added to biology's inventory. Rather, it is a global property of living organization—something that concerns the way processes coordinate with one another across time and space. As early as the 1940s, while working between Princeton and Vienna, Erwin Schrödinger wondered how it was possible for an organism to avoid collapsing into disorder, speaking of an "order that feeds on order." Today we can say that that intuition pointed toward a more precise concept: life is not merely order, but coherent order—capable of sustaining itself, of resonating with itself, and of producing dynamic stability.

It is no coincidence that, in different contexts, scientists and philosophers have used similar imagery. The neurologist Karl Lashley, studying the distribution of functions in the brain in the 1950s, spoke of "field" action, incapable of being localized to a single point. In another field, conductor Wilhelm Furtwängler argued that music does not arise from the sum of its notes, but from the tension that runs through them and unites them. Similarly, biological form does not emerge from the accumulation of parts, but from a coherent relationship among processes, which ensures that the whole takes precedence over its components.

This paragraph thus concludes on a threshold. If form cannot be reduced to either genes or function, if selection is not enough to explain its emergence and stability, then we must change our perspective. We must stop asking ourselves only what is selected, and start asking ourselves how an organism manages to be a meaningful unit over time. The answer that emerges is not a new mechanism, but a new keyword: coherence. This is where the next section will begin, in which life will be viewed not as a series of fortuitous reactions, but as a physical and biological phenomenon capable of maintaining meaning, form, and identity amid the continuous flow of change.

The Myth of the Sovereign Gene: When Information Is Not Enough to Explain Form.

For at least half a century, biology has portrayed life as a story written in DNA. A powerful, elegant, even seductive narrative: four letters, infinite combinations, a molecular book capable of explaining the shape of a leaf, the arrangement of neurons, the complexity of a human organism. It is no coincidence that, in 1962, when Watson, Crick, and Wilkins received the Nobel Prize, enthusiasm spread far beyond the scientific community. In Cambridge, people spoke of the "secret of life" as something that had finally been revealed.

Yet, as the years passed, that promise proved to be partial. Not false—DNA is indispensable—but incomplete. Gene-centrism, that is, the idea that the gene is the fundamental unit sufficient to explain the form and behavior of organisms, began to show cracks precisely where it claimed to be most solid: in development, in the stability of forms, and in the surprising robustness of living things.

One finding, among many, struck the public in the early 2000s: the *"Human Genome Project,"* officially completed in 2003, revealed that humans possess just over 20,000 genes—not many more than a worm like *Caenorhabditis elegans*. The question—embarrassing in its simplicity—began to circulate even outside the laboratories: where, then, is the complexity? If form were truly written entirely in the genes, one would have expected a much more marked difference.

The problem is that DNA does not function like a detailed architectural blueprint. It does not specify where each cell will go, nor how billions of interactions will be coordinated over time. It provides components, possibilities, and chemical constraints. But biological form always emerges within a context, through dynamic relationships that cannot be pinpointed to a single genetic sequence.

An Italian embryologist, during a conference in Pavia in the early 1990s, liked to provoke the audience with a phrase that has remained in circulation among students: *"DNA knows how to make proteins, not organisms*." It was a joke, of course, but it hit the nail on the head. The shape of a heart, the symmetry of a flower, and the arrangement of fingers do not derive from a gene "for the heart" or "for the hand," but from collective, regulatory processes that transcend the individual molecular element.

Here, gene-centrism reveals its deepest conceptual limitation: confusing information with instruction. DNA contains information, but not complete operational instructions. It is like the text of a play without stage directions, without timing, without intonation. Anyone who has seen two different productions of the same Shakespeare play—perhaps a classic at the Globe in London and an experimental one at a Berlin theater—knows how much the final form depends on something that isn't written word for word in the script.

In biology, too, the same genome can give rise to different outcomes depending on the cellular context, mechanical interactions, bioelectrical signals, and the chemical environment. Conrad Waddington's studies on epigenetics, conducted as early as the 1940s in Edinburgh, had already introduced the image of the "epigenetic landscape": a powerful metaphor to convey that development follows preferred trajectories, but ones that are not rigidly predetermined by the gene.

The point, then, is not to downplay the role of DNA, but rather to reframe it. The gene is not a sovereign ruling from above, but rather a tool within a broader system. To continue thinking that form is "contained" within the genome is akin to searching for a melody within individual notes. Life, on the other hand, seems to emerge from a higher level of integration, where information becomes effective only when it is related, synchronized, and made coherent.

This point is crucial to the chapter's argument. It shows why natural selection, acting on genetic variations, cannot by itself explain the emergence of stable and recognizable forms. And it sets the stage for a more radical question: if DNA is not enough, what holds the organism together as a unit of meaning and form? The answer will not be sought in an additional molecule, but in an organizing principle capable of coordinating the many parts. It is here that the discussion of coherence, developed earlier, becomes not only plausible but necessary.

The Gene and the Machine.

Every major scientific era brings with it not only new data but also new dominant metaphors. And metaphors are not mere embellishments of discourse: they guide our perspective, establishing what appears obvious and what becomes unthinkable. Gene-centrism, which has guided biology for decades, did not arise in a vacuum. It is the product of an era that had made technology its horizon of meaning, and technical determinism a sort of implicit philosophy.

It is no coincidence that, starting in the 1950s, DNA came to be described as a code, a program, or an instruction. These were the same years when, between Princeton and Los Alamos, modern computer science was born, and Alan Turing became an almost mythological figure. Biology, fascinated by the explanatory power of the new computing machines, began to speak their language. If a computer works because it follows precise instructions, why shouldn't an organism work the same way?

This analogy had extraordinary power. In the 1970s, as Silicon Valley was taking its first steps and the microprocessor entered the collective imagination, the idea that the gene was the "software of life" seemed not only plausible but inevitable. In a famous lecture in Santa Fe, a molecular biologist recounted that he had come to understand his work "by thinking of the cell as a chemical computer." It was an effective metaphor, but also a conceptual cage.

Technological determinism—the notion that technologies almost automatically drive the evolution of society—has a close relative in biological gene-centrism. In both cases, a single element—the machine or the gene—is elevated to the status of primary cause, while context, relationships, and emerging dynamics take a back seat. It is as if a society could be explained solely by its tools, or an organism solely by its sequences.

As early as the 1960s, the philosopher Lewis Mumford warned against this reductionism, observing that technology is never neutral or self-sufficient, but is always intertwined with values, choices, and ways of life. The same could be said of the gene: it does not act in a vacuum; it does not "decide" anything on its own. Yet the era needed a simple, locatable, controllable principle. The gene perfectly met this symbolic need.

There is a little-known anecdote, recounted by a French sociologist of science at a Paris seminar in 1984. During a visit to a molecular biology laboratory, he noticed that the whiteboards were covered with arrows, linear diagrams, and unidirectional flows. "It looked," he said, "like the control room of a power plant." It wasn't just a way of drawing; it was a way of thinking. The form of the explanation reflected the form of the dominant technologies.

Today, as we talk about networks, distributed systems, and collective intelligence, even the language of biology is slowly beginning to change. The gene is not disappearing, but it is losing its status as an absolute ruler. And this allows us to reinterpret the " " gene-centrism for what it has also been culturally: a powerful, historically situated metaphor, not a definitive truth about life.

Understanding this does not mean relativizing science, but making it more self-aware. Metaphors guide research just as much as data do. Recognizing that DNA was conceived as a machine in an era dominated by machines is a necessary step toward opening our eyes to different images: no longer the gene as a microchip, but the organism as a coordinated process, as an informational structure that stabilizes itself through relationships. It is from this awareness that it becomes possible to imagine—without ideological constraints—a biology that goes beyond determinism, whether genetic or technological.

The Gene on the Front Page: When the Media Turn Biology into Destiny.

There is a precise moment when the gene ceases to be a technical concept, discussed among specialists, and becomes a public figure. This occurs in the second half of the twentieth century, when science becomes a regular feature in newspapers, illustrated magazines, and later on television. It is there that the gene begins to be portrayed no longer as a component of the organism, but as its hidden director, the custodian of a destiny written once and for all.

A symbolic turning point can be traced back to 1962, the year James Watson, Francis Crick, and Maurice Wilkins received the Nobel Prize for their discovery of the structure of DNA. In news reports of the time—from The New York Times to Le Monde—the double helix was often presented as the "key to life." An effective phrase, to be sure, but one already fraught with ambiguity: a key opens a lock; it grants access. The idea that life had finally been "deciphered" was too powerful not to be amplified.

In the 1970s and 1980s, with the explosion of popular science journalism, the narrative became even more pronounced. Headlines such as "Intelligence Gene Discovered," "The

Aggression Gene," and "The Secret of Obesity Lies in DNA" became commonplace. It didn't matter that, in the following paragraphs, the same articles introduced caveats, probabilities, and environmental interactions. The image that remains is that of the gene as the ultimate cause—simple and reassuring in its linearity.

A well-known Italian science journalist recounted, in a 1998 interview, that he had had a lengthy discussion with an editor-in-chief who insisted on titling a piece on the human genome "The Book of Destiny Finally Unsealed." "I know it's not accurate," he told him, "but that's what people understand." That sentence encapsulates the entire tension between scientific accuracy and media logic: the need for a clear story, with a strong protagonist and an understandable ending.

The Human Genome Project, officially announced in 1990 and celebrated in 2000 at the White House by Bill Clinton and Tony Blair, perhaps represents the culmination of this narrative. Television footage spoke of a "complete map of humanity" and an "instruction manual for life." Few remember that, in the same speech, Clinton emphasized that the genome does not determine an individual's destiny. But that detail, in the media frenzy, seemed secondary to the symbolic power of the map and the manual.

This way of framing genetics does not stem from bad faith. It is the result of a structural simplification, typical of modern media, which prioritizes single causes, direct links, and striking images. The gene—invisible yet nameable, microscopic yet conceptually isolable—lends itself perfectly to this role. It is much more difficult to describe distributed processes, dynamic interactions, and forms that emerge over time.

Thus, almost without our noticing it, media discourse has helped to cement an image of life as a pre-written script , in which natural selection is limited to choosing among variations predetermined by DNA. This image has had profound effects not only on the public's perception of biology but also on the way scientists themselves have communicated—and sometimes conceived of—their own work.

Today, as new research highlights the plasticity of development, the importance of cellular and environmental contexts, and the role of global organization, that old narrative is showing all its cracks. Yet its cultural power persists. Understanding how the media have transformed the gene from a possibility into a destiny is not an exercise in external criticism: it is a necessary step toward freeing our perspective and returning to questions about form, about the organism as a coherent unit, and about what in life always transcends a single instruction.

To say "organism" or to say "mechanism": the moral consequences of a metaphor.

Scientific metaphors do not remain confined to books or laboratories. They slip silently into protocols, clinical decisions, and public policies. To say that life is a machine or a coherent process is not merely a linguistic choice: it is an act that shapes research and, above all, the ethics that accompany it.

In the industrial 20th century, thinking of the organism as a machine seemed natural. Machines were the symbol of progress, efficiency, and control. If the body is a device, then illness becomes a malfunction, therapy a repair, and the doctor a specialized technician. This metaphor has produced indisputable results—from modern surgery to pharmacology—but it has also left gray areas that are difficult to ignore.

An English surgeon, practicing in the 1950s at St. Bartholomew's Hospital in London, noted in his diary that *"the patient, under anesthesia, bears a striking resemblance to a watch laid open on the table."* It was a practical observation, not a cynical one. Yet, in that sentence, one can sense the risk of a vision in which the living subject dissolves, giving way to a collection of interchangeable parts.

When life is conceived as a machine, ethics, too, tends to become engineering-oriented: we evaluate what works, what is efficient, and what optimizes a result. It is not surprising that it was precisely within this cultural framework that heated debates on animal and human experimentation developed between the 1960s and 1970s, often framed more in terms of costs and benefits than of meaning and responsibility. The famous 1975 Asilomar conference, dedicated to recombinant DNA, was one of the first attempts to pause and ask not only "Can we do this?" but "What are we treating as an object?"

Thinking of life as a coherent process, however, radically changes the ethical landscape. If the organism is not an assembled machine but an informational structure that sustains itself over time through relationships, then every intervention becomes an act that affects the whole. One cannot "fix" a process without altering its overall balance. Responsibility is no longer merely technical but interpretive.

In the 1980s, the biologist and philosopher Hans Jonas insisted that life possesses an intrinsic vulnerability that calls for protection, not domination. Jonas was not a field biologist, but his ethics of responsibility stemmed precisely from his rejection of a purely mechanical view of living beings. If life is a process, then it also carries an intrinsic value that cannot be reduced to its utility or performance.

This distinction also emerges strongly in the medical field. An Italian oncologist recounted, during a seminar in Milan in the early 2000s, how the way we talk about the disease influences treatment decisions. "If I tell the patient that we must destroy the enemy," he explained, "I enter into a war- al logic. If I speak of restoring a balance, the nature of the alliance that is formed changes." It is not merely a psychological issue: it is a different implicit ontology.

Within the framework of field biology, this distinction takes on even greater significance. Intervening in a stabilized informational structure means recognizing that the organism responds, adapts, and integrates. Ethics can no longer be limited to external rules but must become a focus on coherence. Not everything that is possible is automatically legitimate, because not everything that alters a part respects the process that keeps it alive. After all, the metaphor of the orchestra is more than just a vivid image. In an orchestra, you don't adjust an instrument without listening to the whole ensemble. And no conductor worthy of the name treats musicians like cogs in a machine. Thinking of life as an orchestra—rather than a machine—means adopting an ethic of listening even before one of intervention.

Changing form or giving form?

One of the most persistent misunderstandings in contemporary biology stems from a conceptual overlap that is as widespread as it is rarely articulated: the idea that producing variations is, sooner or later, equivalent to generating organization. After all, it is an understandable temptation. If natural selection acts on random differences, and if time is long enough and the number of trials large enough, why shouldn't form emerge as well?

Yet, there is a gap between change and structure that cannot be bridged by mere accumulation. To change means to alter a state; to organize means to establish relationships that resist change. Life, in its stubborn regularity, belongs to this second category.

A classic example, often discussed in the hallways of evolutionary biology departments, concerns mutations. In the 1970s, at the Natural History Museum in London, a young researcher showed a group of students a series of Drosophila wings deformed by laboratory-induced mutations. "Look at all this variety," he said. Then, pointing to a normal wing: "Now explain to me why this form recurs, always the same, millions of times, despite genetic noise." The question hung in the air. Mutations explained the deviations, not the persistence of the form.

Variation introduces differences, but it does not explain why certain configurations are favored, nor why an organism returns to a stable form even after significant perturbations. A salamander embryo, cut in half during the early stages of development—experiments conducted as early as the late nineteenth century by Hans Driesch in Naples, at the Anton Dohrn Zoological Station—does not produce two defective half-organisms, but two complete individuals. Here the change is drastic, yet the structure reconstitutes itself. Something "knows" what the whole must be like.

This point is crucial: variation is blind to the whole, whereas organization implies a holistic view. Not in the sense of a conscious design, but of a coherence that guides local processes. Classical biology has often confused the two, attributing to selection the credit for creating what, in reality, it merely filters.

An anecdote recounted by a Swedish ecologist during a conference in Uppsala in the early 2000s illustrates this point well. While studying lichen populations subjected to extreme environmental stress, he observed a massive increase in morphological variability. "They changed dramatically," he said, "but they didn't become more complex or more organized. On the contrary, they often lost structure." Variation, pushed beyond a certain threshold, produces disorder, not new form.

The distinction between change and structure is well known outside of biology as well. An architect would say that simply modifying materials is not enough to create a building. A musician knows that varying notes without a harmonic relationship produces noise. It is significant that, in 1948, Norbert Wiener—the father of cybernetics—warned that information is not the same as randomness: "Information is information, neither matter nor energy, but not noise either." A phrase often quoted, but rarely taken seriously in terms of its biological implications.

In a living organism, structure is not a late result of selection, but rather a condition that makes selection possible. Without a relatively stable form, there is nothing to select. Variation acts upon a structure, not in place of it. Confusing these two levels means attributing to time and chance a creative capacity that, empirically, they do not possess.

This linguistic and conceptual confusion has had profound consequences. It has led researchers to seek the origin of form in the mechanisms that explain diversity, rather than in those that ensure coherence. And it has made it difficult to ask what is perhaps the most uncomfortable question: Where does life's tendency to organize itself—to maintain unity through change—come from?

In the context of field biology, this question takes on a new significance. Structure is not seen as an improbable outcome of infinite variations, but as a dynamic attractor—a configuration that the system tends to achieve and preserve. Variation, then, does not disappear, but takes on a different role: it explores the margins; it does not define the center.

Perhaps the true conceptual leap to be made is to recognize that life does not coincide with change, but with that which, while changing, remains. And it is in this fidelity to a form—rather than in the proliferation of differences—that we glimpse an organizing principle still largely to be understood.

The role of attractors in theoretical biology.

For many theoretical biologists, speaking of form without evoking a purpose has always been a delicate exercise. The risk of slipping into finalism—a word that still provokes discomfort in many academic seminars today—has often led to avoiding the problem rather than solving it. Yet, beginning in the 1960s, a concept borrowed from the mathematics of dynamical systems offered an elegant way out: that of the attractor.

An attractor is neither an end in itself nor a hidden plan. It is a configuration toward which a system spontaneously tends, given certain conditions. The term entered biology primarily thanks to scholars such as Stuart Kauffman, who in the 1990s, between Santa Fe and Princeton, began to speak of "*gene regulatory* networks" as systems endowed with stable states. It does not matter how many local perturbations the system undergoes: within certain limits, it always returns there.

The attractor allows us to make a strong statement without saying anything metaphysical. It does not explain why that form exists, but it clarifies how it is maintained. In this sense, it is a valuable conceptual tool for distinguishing between variation and organization. Variation is movement within the space of possibilities; organization is the fact that this space is not flat, but structured by basins of stability.

A developmental biologist recounted, during a summer school in Les Houches in the early 2000s, that he had truly grasped this point by observing a simple model of cell differentiation. By changing parameters and initial conditions, the trajectories shifted, but the final outcomes were surprisingly few. "It doesn't matter where you start," he said, "you always end up in the same place." No purpose, no design: just a dynamic topology that channels the possible.

This is the crucial point: the attractor does not direct, but constrains. It does not tell the system what it must become, but what it is unlikely to become. And in doing so, it restores an active role to form without attributing intentions to it. Form emerges as a global property of the system, not as a prize awarded by selection.

This distinction becomes particularly instructive when looking at computational biology. In recent decades, computer models have shown an extraordinary ability to simulate changes: mutations, fluctuations, population dynamics, rapid adaptations. The animations are elegant, the graphics convincing, the trajectories abundant. Yet, when it comes to reproducing the stability of biological forms, something often eludes us.

An MIT researcher, interviewed informally during a conference in Berlin in 2017, admitted this with a touch of irony: "We're very good at making things happen. Much less so at making them stay." The problem is not technical, but conceptual. Many models are built to explore the space of variations, not to account for the presence of robust attractors.

In the absence of an explicit notion of an attractor, the computer treats form as a contingent result—always reversible, always renegotiable. But in real organisms, this is not the case. A heart does not "try out" infinite configurations before beating; it converges rapidly toward a stable dynamic. An embryo does not explore all possible anatomies; it falls, so to speak, into a very specific valley in the dynamic landscape.

Today, that landscape has become computable, but it often remains uninhabited: models map it out without truly explaining why certain basins are so deep and others nearly invisible.

The consequence is paradoxical: the more we increase computing power, the more we risk confusing computational complexity with biological organization. The computer multiplies the scenarios, but life realizes very few of them. Understanding why—and how—this dynamic selection occurs even before Darwinian selection is one of the open challenges of field biology.

Function comes later: because organisms are not well-designed tools.

There is a mental habit, as deeply rooted as it is rarely questioned, that runs through modern biology: the idea that function comes before form. The heart serves to pump blood, the wing serves to fly, the eye serves to see. From this perspective, form would be little more than an engineering compromise, the result of a long series of adjustments selected for their utility. It is a way of describing life that works well in textbooks, but begins to falter when one tries to take the stability and recurrence of biological forms seriously.

The conceptual reversal proposed here is not a philosophical quirk but an explanatory necessity: it is not function that generates form, but rather form that makes certain functions possible. Functions change, overlap, and readjust; forms, on the other hand, display surprising tenacity. They are like musical scores that allow for many interpretations, but not an infinite number.

An early 20th-century anatomist, Adolf Portmann, observed in Basel that many animal structures are *"richer in form than is necessary for their function.*" He was thinking of the color patterns of deep-sea fish, invisible in the darkness, or the elaborate symmetries of organisms that no predator would ever admire. These were not evolutionary waste products, but clues: form cannot always be justified in terms of immediate utility.

The point becomes even clearer when we look at embryonic development. Before an organ "serves" any purpose, it must take on a coherent, recognizable, stable form. In many species, morphologically similar structures perform different functions; others, with analogous functions, exhibit radically different architectures. In short, function is a belated interpretation of an already established configuration.

An anecdote recounted in the 1970s at the Utrecht Embryology Laboratory concerns a series of experiments on slightly deformed amphibian embryos. The cells, even when displaced, "knew" where to go. Not because they knew the final purpose, but because the organism's overall form served as a constraint, an organizational framework within which cellular behaviors found meaning. Function emerged as a consequence, not as a guide.

This reversal challenges a certain simplified Darwinian narrative, according to which natural selection sculpts forms based on advantageous functions. Selection can refine, eliminate, and stabilize; but it works on forms that are already available—it does not invent them from scratch. As paleontologist Stephen Jay Gould noted during a lecture at the American Museum of Natural History in New York in 1984, *"evolution is a great recycler, not a designer from scratch.*" A structure that arose in a certain formal context can be co-opted for entirely new functions, without its fundamental architecture being redesigned.

Thinking of form as a priority also means ceasing to treat the organism as a technical object. A knife is defined by its function; if it doesn't cut, it's a bad knife. An organism is not. A lung is not a device for gas exchange: it is a living configuration that makes that exchange possible, along with many other side effects—some of which are biologically irrelevant, others decisive.

This shift in perspective has profound consequences. It shifts the focus from the question "What is it for?" to the question "What kind of form is possible here?" And it forces biology to confront an uncomfortable fact: life does not optimize functions; it stabilizes configurations. Function is a narrative we construct afterward—often useful, sometimes illuminating, but rarely foundational.

It is no coincidence that architects like Frei Otto, while studying lightweight structures in Stuttgart in the 1960s, drew more inspiration from biological forms than from their functions. Networks of soap bubbles, the hollow bones of birds, and the patterns of leaves revealed an autonomous formal logic, capable of generating efficiency without being designed for a specific purpose. Biology, in this context, does not teach us what to do, but how forms emerge.

To say that form precedes function does not mean denying adaptation, but rather putting it into perspective. It means recognizing that life does not arise to be useful, but to be coherent. Function is a local consequence; form, a global decision of the system.

Same architecture, a thousand uses: when form does not dictate destiny.

If form precedes function, as the conceptual shift in this chapter suggests, then we must explain a fact that is only seemingly paradoxical: extraordinarily stable biological forms can support extremely variable functions. It is here that biology ceases to resemble an engineering textbook and begins to tell a more nuanced story, one of repurposing, deviations, and controlled improvisations.

Neurobiology offers examples that are hard to ignore. In the 1960s, at MIT, Paul Bach-y-Rita observed that blind patients could "see" through tactile stimulation on their backs or tongues. The visual cortex, deprived of ocular input, did not lose its shape or organization, but its function c d: it processed information from other sensory channels. It was not the function that defined the brain area, but rather its internal architecture, capable of accommodating different contents while maintaining structural coherence.

A similar phenomenon is found in comparative physiology. The lungs of terrestrial vertebrates and the swim bladder of bony fish share a clear morphological relationship. Yet the same basic form lends itself to radically different functions: aerial respiration, control of hydrostatic buoyancy, and sound production. As the Swedish physiologist Per Scholander noted during an Arctic expedition in the 1950s, form is not a functional destiny, but an open resource.

Even more instructive is the case of the heart. In school textbooks, it is a pump; in marine mammals, such as the Weddell seals studied in Antarctica since the 1970s, it also becomes a sophisticated metabolic modulator, capable of drastically slowing its beat during deep dives without changing its fundamental structure. The heart's form remains recognizable, almost conservative; its function, however, adapts to extreme physiological conditions.

These examples show that stability of form does not equate to functional rigidity. On the contrary, it is precisely this stability that makes plasticity possible. A coherent form acts as a field of possibilities: it does not prescribe a single use, but makes many uses feasible. The French biologist Alain Prochiantz, speaking at a conference in Paris in the early 2000s, observed that the brain "is not plastic because it is indeterminate, but because it is deeply structured."

Here, natural selection appears in a different light. It does not select isolated functions, but rather forms capable of withstanding functional variations without collapsing. Survival does not favor the most specialized organ, but rather the most reusable one. This logic is closer to language than to a machine: a stable grammar allows for ever-new sentences.

An anecdote recounted by an Italian physiologist concerns a laboratory in Naples in the 1990s, where cave-dwelling animal were studied. Some fish, lacking functioning eyes, nevertheless retained their orbital structures. These no longer served for seeing, but for organizing alternative sensory circuits. The form remained as a memory of possibility, not as a useless remnant.

The functional plasticity of stable forms subtly suggests a powerful idea: life does not solve problems once and for all, but builds structures that remain open to interpretation. Function is a contingent response; form, a lasting context. And it is precisely this asymmetry that makes organisms resilient, creative, and historical.

Because nature appears beautiful to us before it appears useful.

There is a moment, when faced with a living organism, when the mind falls silent and only the gaze remains. It happens when observing the spiral of a seashell, the unsettling symmetry of a jellyfish, or the sharp outline of a ginkgo leaf, which has remained virtually unchanged for over two hundred million years. In that instant, we are neither interpreting nor deciphering symbols: we are recognizing a form. And it is precisely this immediate recognition that challenges the idea that form is merely a functional byproduct of natural selection.

Evolutionary biology explains well why a structure functions, but says less about why a form appears necessary, obvious, almost inevitable. The Swiss naturalist Adolf Portmann, writing in the 1950s after extensive field observations along the Adriatic coast, noted that many animals exhibit visible traits that do not seem reducible to either survival or reproduction. His question, then as now, remained unanswered: to whom do the forms of life reveal themselves?

This question is not far removed from the one artists ask themselves. Paul Cézanne, working obsessively on Mount Sainte-Victo between 1882 and 1906, said he wanted to "make the structure visible." Not meaning, not emotion, but form as internal tension, as an order that manifests itself. Cézanne did not paint symbols of nature: he sought its formal coherence, something that precedes any interpretation. It is striking how much this quest resonates with the biological idea of form as an autonomous organizing principle.

Even the philosophy of form, when at its most restrained, avoids symbolism. Maurice Merleau-Ponty, in his lectures at the Collège de France, insisted that perception is not a reading of the world, but a participation in its structure. We see forms because we ourselves are living forms. The stability of a biological morphology strikes us not because it "means" something, but because it resonates with our own way of being organized.

Form, in this sense, is not a message to be deciphered, but a perceptual event. It does not refer to anything else: it simply is. And this is perhaps why biological forms have always attracted architects, designers, and sculptors. When Frei Otto was studying lightweight structures in Stuttgart in the 1970s, he observed soap bubbles and cellular networks not to copy them symbolically, but to understand how matter, left to its own devices, finds stable configurations. The living form thus becomes a lesson in coherence, not a repertoire of metaphors.

From this perspective, the priority of form over function also acquires cognitive value. We recognize life because we recognize its forms, even before we know what they do. An embryo, a cell, an unknown organism: we identify them as living beings by the way they occupy space, by their perceptible internal organization. Aesthetics, here, is not a cultural luxury, but a primary mode of access to reality.

Natural selection, then, truly resembles "the blind man and the orchestra": it can refine, eliminate, and adapt, but it does not explain why certain forms resonate, persist, and impose themselves on the human gaze across different eras and cultures. Form is not an ornament of life; it is the way life appears to the world.

Showcases of Form: When Museums Taught Biology to See.

Entering a major nineteenth-century natural history collection was not merely a scientific experience. It was, first and foremost, an exercise in seeing. The halls of *the "Muséum national d'Histoire* naturelle" in Paris, reorganized after the Revolution, or those of *the "Kaiserlich-Königliches Naturhistorisches* Hofmuseum" in Vienna, inaugurated in 1889, did not merely preserve specimens: they staged form. Rows of skeletons, series of shells, and taxidermied birds in almost theatrical poses composed a sort of grand visual score. And it was precisely here that Darwinism, though triumphant as a theory of adaptation, began to reveal a gray area: form—stable, recognizable, almost obstinate—seemed to precede function.

The naturalists of the time were well aware of this, even if they often did not say so openly. Georges Cuvier, who built one of Europe's most influential comparative collections in Paris, was convinced that every organism was a coherent whole, governed by a "correlation of parts." There is a famous anecdote—reported by several of his students—according to which Cuvier, upon observing a single bone, claimed to be able to reconstruct the entire animal. Hyperbole aside, the point was clear: form was not a side effect of adaptation, but an organizing principle. Darwin, who admired Cuvier, never quite managed to integrate this insight into the mechanism of natural selection.

Nineteenth-century collections functioned as enormous cognitive devices. Sorting, juxtaposing, and comparing meant training the eye to discern patterns, recurrences, " " and variations on a single theme. In Vienna, in the halls dedicated to vertebrates, the sequences of skulls displayed a formal continuity that spanned different species. It was not difficult for an attentive visitor to perceive that variation does not coincide with organization. One can change a great deal without losing a form; and one can lose a form without changing much. It is a subtle but decisive distinction.

Charles Darwin visited museums and collections several times during his European travels, and in his letters—for example, to Joseph Dalton Hooker in the 1860s—a certain unease regarding the "tyranny of forms" shines through. Natural selection explained why a structure was advantageous, but not why that particular structure emerged rather than another. The display cases, silent yet insistent, seemed to suggest that life followed deeper scores, like an orchestra that plays even without seeing the conductor.

In this sense, the collections served as schools of aesthetics as well as of science. Ernst Haeckel, a German zoologist and tireless museum-goer, transformed that visual experience into art with his **Kunstformen der* Natur* (1899–1904). The plates dedicated to radiolarians or jellyfish were not mere illustrations: they were philosophical statements. For Haeckel, form possessed ontological dignity. It was not an epiphenomenon, but the visible expression of an internal order. That Haeckel often stretched the data to fit his theories is another matter; the fact remains that his perspective had been shaped in the halls of museums.

In Paris, Étienne Geoffroy Saint-Hilaire—Cuvier's longtime rival—advocated the idea of a "plan of organization" common to all animals. In the famous 1830 debate *at the Académie des Sciences*, the question was not merely evolutionary but formal: does a structure exist that precedes functions? The collections, with their orderly series, seemed to support Geoffroy's view at least as much as Cuvier's. Walking among the display cases, visitors did not see isolated adaptations, but families of forms—variations on a morphological theme that withstood the test of time and the environment.

It is no coincidence that many nineteenth-century biologists were also excellent draftsmen. Before the advent of scientific photography, the hand and the eye worked together. In Vienna, Carl Gegenbaur taught comparative anatomy by having students obsessively draw the same structures in different species. The exercise served not only to memorize, but to "feel" the form.

One of his students, in a diary now preserved in an archive, recounts his initial frustration and then his surprise: *"At a certain point,"* he writes, *"I no longer saw bones, but relationships."*

This training of the eye had theoretical implications. If form takes precedence, then function becomes a local modulation, not the generative principle. Natural selection does act, yes, but as a filter, not as an architect. The nineteenth-century collections, without making any grand proclamations, subtly suggested this idea. They showed that life does not proceed through random additions, but through the stabilization of configurations. In contemporary terms, we might say: informational structures that remain coherent amid the flow of change.

Today, accustomed as we are to thinking in terms of genes, nineteenth-century collections continue to speak to us. Not as relics of outdated knowledge, but as archives of a question that remains open: where does form come from? Darwinism answers the "why" of adaptation; museum display cases, silently, remind us that form is an autonomous problem, an event of order rather than utility.

Scientific photography and the rediscovery of living form.

There was a moment in the twentieth century when biology stopped drawing and began photographing. It was not merely a change in technique, but a genuine shift in perspective. Scientific photography—from optical microscopy to electronic , and on to advanced cellular imaging—has radically transformed the way we perceive living form. Without resorting to symbols, metaphors, or interpretive frameworks imposed from the outside, images have begun to show life as it unfolds, revealing structures that cannot be reduced to either function or the gene.

At the turn of the century, in laboratories in Berlin and Jena, the first cellular microphotographs produced an almost unsettling effect. Cells did not appear as shapeless sacs of chemical reactions, but as organized spaces, traversed by tensions, boundaries, and symmetries. Walther Flemming, who had already described mitosis in the late nineteenth century, realized—as he observed increasingly sharp slides—that cell division was not a random event driven by individual molecules, but a highly orchestrated sequence of transformations in form. Photography made evident what theory still struggled to articulate: variation is not organization.

With the advent of electron microscopy in the 1930s and 1940s, particularly in the laboratories of Toronto and Cambridge, the cell definitively shed its "black box" image. Images of mitochondria, endoplasmic reticulum, and layered membranes showed that function arises from a precise and stable spatial arrangement. Albert Claude, who was awarded the Nobel Prize in 1974, recounted in an interview that he felt a sort of vertigo the first time he saw the interior of a cell under an electron microscope: "I *wasn't looking for machines*," he said, "*but* for *architecture*." A simple, yet revealing statement.

It is here that Darwinism, understood in a strictly adaptive sense, encounters a conceptual limit. Natural selection can explain why a structure is preserved, but scientific imagery reveals just how coherent that structure already is prior to any measurable functional advantage. Images do not "prove" a theory, but they shape our perception: form appears as a primary given, not as a byproduct. The blind man—to return to the metaphor from the " " chapter—can still speak of selection; but the orchestra, finally visible, reveals a score.

In the 1950s and 1960s, time-lapse photography—introduced by Jean Comandon in France and later perfected in American laboratories—made it possible to observe living cells over time. No longer static snapshots, but continuous processes: cell migration, tissue self-organization, and embryonic development cycles. Watching these films, many biologists realized that change does not destroy form, but preserves it through controlled transformations. Structure is not rigid, but persistent. A concept difficult to capture in genetic terms, but immediately evident to the eye.

In the 1980s and 1990s, with fluorescence and *"confocal imaging*," form ceased to be merely visible and became readable as spatial information. Labeled proteins did not appear randomly distributed but followed recurring patterns, concentration fields, and stable gradients.

These images also had a broader cultural impact. Artists such as Hiroshi Sugimoto or, in the scientific realm, Lennart Nilsson—famous for his embryonic photographs published in *Life* in 1965—showed the general public that life is not a sequence of fragments, but a continuum of forms that stabilize over time. Nilsson, while working in popular science contexts, insisted that his images were not "interpretations" but technical revelations. The form spoke for itself.

Twentieth-century scientific photography, therefore, did not simply add details to the biological picture: it challenged gene-centrism by showing that DNA, while fundamental, is not sufficient to explain the visible order of life. The images suggest that biological information is not merely encoded, but distributed across space and time. Form, far from being a mere ornament, appears to be the true locus where information stabilizes.

In this sense, photography has restored to biology something it risked losing: direct contact with the ph . Without symbols, without allegories, without premature reductions. Only light, matter, and organization. And perhaps it is precisely from here that a broader reflection on life as a field can begin anew: not as a sum of elements, but as a coherent event that reveals itself even before it can be explained.

Natural aesthetics as a form of literacy in life.

There comes a moment in the study of life when the explanation arrives too soon. Before even asking what a form is for, one should learn to see it. It is a step often overlooked in modern scientific education, yet crucial for understanding what natural selection alone cannot explain: the emergence, coherence, and stability of biological forms. Natural aesthetics, in this sense, is not a cultural ornament but a genuine education in perception.

In his **Italian* Journey*, Goethe recounts how the direct study of Mediterranean plants in Palermo in 1787 taught him more about living forms than he had learned from books. He was not seeking functions or adaptive advantages, but rather relationships: variations of the same structure, recognizable metamorphoses, an order that manifests itself through differences. His *Morphology* arises precisely from this discipline of the gaze, from a patient attention to form as a primary phenomenon.

This kind of visual education is not immediate. Alexander von Humboldt, in his South American field notes from 1799 to 1804, observed that the inexperienced naturalist sees only “a jumble of green,” while those who have trained their eyes distinguish rhythms, symmetries, and recurring patterns. It is not yet experimental science, but neither is it a subjective impression: it is the learning of form as an objective, shareable, and recognizable reality.

This is where Darwin’s problem of form comes in. Natural selection explains why a structure is retained or discarded, but it does not teach us how to recognize it. Without a trained perception of living configurations, biology risks being reduced to a list of functions and advantages, losing sight of what is actually selected: an already organized form. It is like trying to explain the success of a building by talking only about the materials, without ever considering the architecture.

In the 19th century, before biology became predominantly experimental, aesthetic observation was an integral part of the method. John Ruskin, whom we remember today primarily as an art critic, taught botanical drawing courses at Oxford in the 1870s. Not to train artists, but to teach people how to see. He argued that drawing a leaf forced one to grasp its structure, not its utility. This idea was shared, during those same years, by some comparative anatomists who saw drawing as a means of knowledge, not embellishment.

This perspective, trained to perceive form, is what is missing when life is reduced to genetic information or adaptive function. Form is not a byproduct, but an organizing principle that demands recognition before it can be explained. Field biology emphasizes this point: the organism is a stabilized informational structure, and stability manifests itself first and foremost as a perceptible configuration.

An anecdote told by an elderly zoology professor at the University of Padua illustrates the point well. During an exam, he would show students a series of marine organisms without asking for their names or functions. The question was always the same: "What do you see recurring?" Those who answered by speaking of symmetries, proportions, and formal continuities passed the exam. Those who spoke only of adaptations did not. *"Form first, explanation second,"* he would say.

From this perspective, natural aesthetics is not a humanistic luxury, but a cognitive skill. It is the ability to grasp order before meaning, structure before function. Just as happens when learning a language: first you recognize the sounds and rhythms, and only later do you understand the meaning of the words. Life, as the language of nature, demands the same kind of attention.

The Stability of the Living: Why Form Demands a Deeper Theory.

There is a point in biological reasoning where form ceases to be a descriptive issue and becomes a theoretical question. As long as we limit ourselves to saying that a structure "works," we remain within the reassuring realm of adaptation. But when we ask why that structure holds together—why it persists over time despite continuous molecular turnover—then Darwinism reveals its most subtle limitation. Not because it is false, but because it is not enough.

The priority of form, at the heart of this chapter, naturally paves the way for the subsequent chapters. If form cannot be reduced to either the gene or function, then there must be a principle that maintains it, makes it recognizable, and protects it from dissolving into the flow of variations. That principle, which in the following chapters will be called coherence, still appears here as a necessary intuition.

A classic example, often cited in early 20th-century embryology courses, is that of Hans Spemann's experiments in Freiburg, which culminated in 1924 with the discovery of the "organizer." By transplanting a small embryonic region into another newt embryo, Spemann observed the formation of a complete body axis. Not a mere sum of cells, but a coherent organism. Form emerged as an ordered whole, capable of reorganizing itself. Natural selection can explain why such an organized organism is successful; it does not explain why the organization arises and is maintained with such robustness.

This is where the problem of form becomes a conceptual bridge. The stability of biological structures is not static but dynamic. Cells change, molecules are renewed, and yet the organism remains itself. Form, then, is not a frozen design but a configuration that persists because its parts hold together according to internal rules. It is the same insight that led the physiologist Walter Cannon, at Harvard in the 1930s, to introduce the concept of homeostasis: not as immobility, but as the ability to maintain an overall configuration. Once again, function explains how a disturbance is responded to; form suggests that a higher level of integration exists.

The blind man and the orchestra come in handy here as well. Natural selection listens to the final result—survival—but does not "see" the score that holds the musicians together.

Recognizing the priority of form means admitting that, prior to adaptation, there is a coherence that makes any adaptation possible. It is not a mysterious entity, but an emergent property of living organization.

Thus, the chapter on form does not close in on itself. On the contrary, it sets the stage. If life shows a surprising fidelity to itself amid change, then we must shift our gaze: from local causes to global properties, from isolated mechanisms to the relationships that coordinate them. It is this shift that naturally leads to the idea of coherence as a fundamental property of the living, the central theme of the next part of the book. Form, ultimately, is the first visible clue to an invisible yet active coherence. Understanding it does not yet mean explaining it fully, but rather recognizing that life is not merely what reacts, but rather what remains held together.

PART TWO. The Invisible Thread of Life.

From the thought of Erwin Schrödinger.
Life is not defined by what it is made of, but by its ability to maintain an improbable order over time.

From the thought of Ludwig von Bertalanffy.
An organism is not the sum of its parts, but a system whose properties emerge from the set of relationships.

From the thoughts of Ilya Prigogine.
Living order is not static: it exists only as a process that continually renews itself far from equilibrium.

From the thoughts of Francisco Varela.
Understanding life means understanding how a system produces and maintains its own unity.

After challenging the classical biological paradigm, this second part of the book takes a decisive step: it shifts the fundamental question from composition to stability. While the first part clearly demonstrated that life cannot be explained as a simple chain of reactions or as the blind result of selection, here the problem becomes more radical and, at the same time, more concrete: what allows an organism to remain itself amid continuous change?

The proposed answer is neither a new mechanism nor an additional force to be added to those already known. It is a shift in perspective. Coherence is introduced as the fundamental property of living beings: not a technical detail borrowed from physics, but the principle that makes the dynamic unity of organisms possible. To live, from this perspective, means maintaining a recognizable overall form while everything, at the level of the parts, flows and transforms.

The third chapter clarifies the precise meaning of this statement. Coherence is not a evocative metaphor, nor is it a vague concept: it denotes a specific physical condition—already known in other scientific fields—that takes on decisive significance in biology. Life thus appears as an active process of resistance to fragmentation, a continuous negotiation with the natural tendency toward dispersion. There is nothing mysterious about this emergence: rather, there is a level of organization that requires appropriate conceptual tools to be described.

Chapter Four extends this insight to its deepest implications. If coherence is a global property, then the organism cannot be conceived of as a machine assembled piece by piece. Biological causality is neither linear nor localized: the parts do not "decide" on their own, but behave according to the overall state of the system. The whole, in other words, is not the final result of the functioning of the parts; it is their condition.

This part of the book therefore occupies a central position in the overall logical progression. It fills the void left by the critical " " to reductionism and prepares the ground for the introduction of biological fields, which will be addressed in the next section. Without the concept of coherence, the idea of a field would remain abstract; without the primacy of the organism, information would remain disembodied.

Here, biology begins to speak a different language: no longer that of the accumulation of local causes, but that of the relationships that hold living beings together. It is a theoretical shift, but also a cultural one. For recognizing coherence as the true name of life means accepting that unity is not an illusion to be dismantled, but rather a real phenomenon to be understood.

III. Coherence: The True Name of Life.

There is a point—often overlooked—at which biology ceases to be a catalog of molecules and becomes a reflection on the form, meaning, and endurance of life over time. This is where coherence comes into play. Not as an evocative word or poetic metaphor, but as an operational, measurable principle capable of bridging physics and biology without losing its rigor.

This chapter stems from a simple yet unsettling observation: life cannot be explained as the mere sum of local processes, because what defines it is the ability to maintain a dynamic unity under conditions that, for any inert system, would quickly lead to disintegration. To live means to be together, to remain in harmony, to continue "being one" while everything pushes toward dispersion.

In physics, coherence refers to a very specific condition: the stable correlation between the parts of a system, such that the behavior of the whole cannot be reduced to the behavior of the individual isolated components. This is not a vague idea. It is at the heart of rock-solid experimental phenomena, from quantum superposition to phase coherence, and represents one of the most profound conceptual achievements of twentieth-century science. When this concept is applied to biology, it does so not as an exotic import, but as a surprisingly apt interpretive key.

The thesis running through the chapter is clear: life is a process of actively maintaining coherence. An organism lives as long as it manages to preserve a functional correlation among its parts, among its levels, and between what happens locally and the global order that emerges from it. In this sense, to live is to resist decoherence—that is, the loss of connection, the fragmentation of information, and the breakdown of dynamic unity.

This idea does not introduce any additional mystery into science. On the contrary, it restores intelligibility to phenomena that reductionism struggles to explain: the stability of biological forms, the speed of integrated responses, and the surprising robustness of organisms in the face of continuous perturbations. Coherence is not a hidden entity, but a real emergent property, observable in its effects and describable using well-established theoretical tools.

It is no coincidence that some of science's greatest thinkers have pointed, in different words, toward this very direction. Erwin Schrödinger, in his famous essay *"What Is Life?"*, wrote that *"the living organism feeds on order,"* anticipating the idea that life does not merely dissipate energy but uses it to maintain a coherent structure over time. Years later, Ilya Prigogine would clarify that *"life is possible only far from equilibrium,"* identifying this very maintained—and governed—distance as the condition for dynamic coherence. And David Bohm, reflecting on the relationship between physics and totality, observed that *"in a universe based on fragmentation, no lasting meaning can emerge,"* a statement that, when interpreted from a biological perspective, sounds like a programmatic declaration.

Within the overall framework of the book, this chapter plays a crucial role. It serves as the conceptual bridge between the crisis of the classical biological paradigm—analyzed in the first part—and the introduction of biological fields as stabilized informational structures, which will be developed in subsequent chapters. Without a clear understanding of coherence, the concept of a field would remain merely a suggestion; with it, it becomes a theoretical necessity.

Here it is made clear that speaking of coherence does not mean denying biochemistry or genetics, but rather placing them within a broader framework, in which molecules matter because they are coordinated, not merely present. Coherence is what allows the whole to precede the parts—not in a chronological sense, but in a logical and functional one. First comes the dynamic unity of the organism, then the details of its mechanisms.

In this sense, the chapter prepares the reader for a shift in perspective: from life as a machine to life as an integrated process of physical meaning. It is not a matter of abandoning science, but of following it to its most radical consequences.

Living Time: How Coherence Creates Organisms' Internal Clocks.

One of the most fascinating aspects of biological coherence emerges when viewed from the perspective of time. Life, in fact, does not simply inhabit the physical time measured by clocks, but produces its own internal temporality, qualitatively different from the uniform succession of instants that governs inanimate systems. A coherent organism does not "endure" time: it structures it, modulates it, and makes it functional to its own organization.

In classical physics, time is an external variable, identical for all processes. In biology, however, time becomes an emergent property of the living system. Every organism maintains a plurality of coordinated rhythms—metabolic, cellular, physiological—that do not necessarily coincide with environmental time but must remain mutually coherent to ensure the stability of the whole. When this temporal coherence breaks down, the organism does not simply "age": it loses its integration.

It is no coincidence that as early as the first decades of the twentieth century, Jakob von Uexküll spoke of the time specific to organisms, observing how every form of life constructs a specific relationship with duration, anticipation, and simultaneity. Later, the theoretical biologist Francisco Varela would clarify that *"biological time is not a framework, but a result of living organization,"* a formulation that fits perfectly into the discourse on coherence. A living system exists as long as it manages to keep its internal processes synchronized, creating an operational continuity that does not coincide with mere material persistence.

An eloquent example comes from chronobiology: circadian rhythms are not simply passive responses to sunlight, but endogenous oscillations that are maintained even in the absence of external stimuli. This means that the organism possesses an internal time that must remain attuned to—but not identical with—the time of the environment. Coherence, in this context, manifests as the ability to hold together the biological past (memory of previous conditions), the operational present, and anticipation of the immediate future. From this perspective, living is a continuous act of synchronization.

Henri Bergson, though operating within a philosophical framework, had grasped this distinction with great clarity when he distinguished the spatialized time of science from lived *"durée."* Today, that insight finds scientific resonance: biological duration cannot be measured solely in quantitative units, because it is qualitatively structured by the coherence of the system. A healthy organism is not one that "lasts the longest," but one that maintains an internal continuity of meaning among its processes.

From this perspective, coherence guarantees not only the organism's spatial order but also its temporal stability. Life continues as long as the system manages to renew itself while remaining true to itself—that is, as long as the flow of change does not sever the thread connecting its transformations. Biological time is not a straight line: it is a fabric of interwoven rhythms, continually reconstructed.

Imperfect Order: Why Life Needs Noise.

There is a persistent misconception when discussing biological coherence: the idea that it coincides with a rigid, silent, disturbance-free order. Nothing could be further from the reality of living beings. Organisms do not eliminate noise; they inhabit it. Life is not a perfect crystal, but a system that remains coherent despite being permeated by continuous fluctuations, microscopic uncertainties, and spontaneous variations. It is precisely this ability to integrate disorder within functional limits that distinguishes biological coherence from any mechanical order.

In physics and engineering, noise is traditionally considered a problem: something to be filtered out, attenuated, or suppressed. In biology, however, noise is inevitable and, in many cases, necessary. At the molecular level, gene expression is inherently noisy: even identical cells, with the same genetic makeup and in the same environment, produce different amounts of proteins. For decades, this phenomenon was viewed as a flaw in the system. Today we know that this variability is a resource: it allows for flexibility, adaptation, and the exploration of possible states without losing the organism's overall identity.

A story often recounted in systems biology laboratories concerns experiments conducted at Stanford in the early 2000s, when researchers began measuring the variability of gene expression in individual cells in real time. The researchers were surprised not so much by the existence of noise as by the fact that it never degenerated into chaos. The fluctuations were real, but they remained confined within precise limits, as if the system knew just how far it could afford to oscillate. Coherence, in this context, is not the absence of noise, but the control of noise.

This principle finds an elegant parallel in the theory of stochastic resonance, developed in the 1980s by Roberto Benzi and colleagues while studying physical and climatic systems. In certain contexts, a moderate amount of noise does not disturb the signal—it amplifies it. Something similar happens in biology. Fluctuations allow living systems to respond to weak stimuli, to avoid becoming trapped in overly rigid states, and to remain sensitive without losing stability. Vital coherence is, in this sense, an elastic coherence.

Hermann Haken, a German physicist and founder of synergetics, liked to say that *"order arises from disorder, but not all disorder generates order.*" Organisms seem to have solved precisely this problem: they select, absorb, and redistribute fluctuations in a way that sustains their overall form. They do not fight instability; they use it as informational fuel. An organism completely devoid of noise would be incapable of adapting, learning, or even developing.

A striking example comes from embryonic development. During the early stages, cells exhibit highly variable behavior, yet the final result is surprisingly robust: properly formed organs, reliable proportions, and coherent architectures. As biologist Lewis Wolpert noted in his studies on morphogenesis, the final precision emerges from imprecise local processes, coordinated by global constraints. Once again, coherence does not eliminate error: it channels it.

This ability to integrate noise profoundly distinguishes living systems from machines. A machine functions well only as long as perturbations are minimal. A organism, on the other hand, functions precisely because it is continually perturbed. Biological coherence is not fragile; it is resilient. It does not amount to total control, but rather to a distributed regulation that tolerates uncertainty without losing direction.

In this light, life appears as a unique form of dynamic order: an order that breathes, oscillates, makes slight errors, but does not fall apart. Noise is not the enemy of coherence; it is its permanent test bed. And as long as the organism manages to transform fluctuations into meaningful variations without losing unity, life continues.

Who governs whom? The decisive difference between automation and living organization.

When discussing biological coherence, the risk of slipping into a mechanistic misunderstanding is always lurking. The idea that an organism "regulates itself" is often translated, almost automatically, into the language of control: sensors, commands, responses. But living regulation is not top-down control, nor is it a program that issues orders to passive components. It is a distributed process, without a command center, in which coherence emerges from the continuous interaction of the parts.

The comparison with machines is instructive precisely because of what doesn't add up. A thermostat maintains a room's temperature by correcting a deviation from a preset value. It works well, but only under limited and predictable conditions. An organism, on the other hand, does not have a fixed set point: it continually redefines what constitutes equilibrium, based on its history, internal state, and context. As Walter Bradford Cannon observed—who introduced the concept of homeostasis at Harvard in the 1930s—biological stability is not immobility, but "stability through change."

The crucial difference lies here: mechanical control reduces variability, while living coherence organizes it. In an automatic, " " system, error is something to be eliminated; in an organism, deviation is information. It is what allows for learning, compensation, and the reorientation of action. A heart does not "obey" a central command ordering it to speed up: it responds to a network of chemical, mechanical, and nervous signals that have no conductor but remain coherent with one another.

In the 1950s, the British physician and mathematician W. Ross Ashby formulated the *"law of requisite variety*": only a system with sufficient internal variety can effectively regulate a complex environment. Applied to biology, this idea clarifies why life cannot be reduced to a control algorithm. An organism is regulatory because it is rich in possibilities, not because it is rigidly programmed. Its coherence does not stem from the imposition of rules, but from the ability to hold together multiple degrees of freedom.

An anecdote told by an Italian physiologist who worked in Pavia in the 1970s illustrates the difference well. During a demonstration, he showed students an automatic device that maintained a constant concentration in a solution. He then compared the experiment to a cell culture subjected to unexpected environmental variations. The device failed as soon as conditions changed too much; the cells did not. He commented, *"The machine corrects itself; the cell reorganizes itself.*" It is a subtle but decisive distinction.

Norbert Wiener, the father of cybernetics, was also aware of this limitation. Although he provided fundamental tools for understanding regulation, he cautioned that living systems could not be equated without reservation with automatic machines. Biological coherence is not a control mechanism that blocks change, but a dynamic that keeps the system open, capable of responding to the unexpected without losing its identity.

Clarifying this distinction is essential for understanding the biology of fields. If coherence is confused with control, the biological field becomes a hidden software that directs matter. If, on the other hand, we recognize that coherence is an emergent property of the relationship between parts, then the field appears, by , for what it is: a stabilized informational structure, not a command mechanism. Life is not remote-controlled. It holds itself together.

What does "coherence" really mean?

In everyday language, the word "coherence" evokes something moral or narrative: being coherent means not contradicting oneself, staying the course. But in physics, the term has a much more precise, technical, and surprisingly concrete meaning. It is not a figure of speech, nor a linguistic quirk imported to lend an aura of depth to biology. Coherence is a measurable property of physical systems, with observable effects and everyday applications, from lasers in hospitals to quantum chips in laboratories.

To understand what we're talking about, we need to take a step back to the time when 20th-century physics began to challenge the classical idea of a world made up of separate objects that interact only locally. In 1927, at the fifth Solvay Conference in Brussels, while Einstein, Bohr, and Schrödinger were engaged in a heated debate over the meaning of quantum mechanics, a concept emerged that was destined to become central: particles are not just "things," but also waves, and waves can be coherent or incoherent.

Strictly speaking, coherence describes the degree of phase correlation between different parts of a wave system. Two waves are coherent when they oscillate in a coordinated manner, maintaining a stable relationship over time. If this relationship is lost, the process is called decoherence. This is no minor detail: without coherence, the laser would not exist. The laser beam—so different from the light of a light bulb—is light in which billions of photons oscillate "in unison." It is an almost textbook example of " r collective order": no longer random emissions, but synchronized behavior.

Charles Townes, one of the fathers of the laser, recounted that the idea came to him in 1951, while sitting on a bench in Washington at six in the morning, as he sought a way to "force" atoms to emit light in a coordinated manner. He wasn't looking for a metaphor: he was pursuing a precise physical condition. When coherence emerges, the system does something qualitatively new.

This is the crucial point: coherence is not about individual elements taken one by one, but about the way they come together. It is a property of the system as a whole. In condensed matter physics, the same principle explains superconductivity: below a certain temperature, electrons in a metal cease to behave as independent entities and form a collective coherent state. The result? Electric current flowing without resistance. A phenomenon so counterintuitive that, when Heike Kamerlingh Onnes first observed it in 1911 in Leiden, he himself had trouble believing it.

For a long time, coherence remained confined to these "exotic" fields: optics, quantum physics, and low-temperature states of matter. But at the beginning of the 21st century, something changed. In 2007, a group of researchers led by Graham Fleming at the University of California, Berkeley, published an unexpected finding in **Nature**: in certain photosynthetic complexes, at biological temperatures, energy appears to be transferred via coherent states. It wasn't definitive proof of *"quantum biology,"* but it was enough to force many to reconsider a preconception: coherence isn't as fragile as once thought, nor is it confined to cryogenic laboratories.

It is important to be clear here. Speaking of coherence in biology does not mean that cells "think quantum-mechanically" or that life violates the laws of classical physics. It means recognizing that physics offers concepts to describe forms of dynamic order that cannot be reduced to the sum of their parts. Coherence is one such concept, and it is scientifically legitimate because it is definable, observable, and subject to experimental verification.

A good, often overlooked example comes from the physics of complex systems. When a set of oscillators—such as pendulums, artificial neurons, or molecules—synchronizes, we are observing a phase transition, similar to those that occur when water freezes or a metal becomes magnetic.

At first, each oscillator follows its own rhythm, in a disorderly manner. But when the interaction between them exceeds a certain threshold—for example, by increasing the strength of the connection or reducing noise—the system's collective behavior changes. Suddenly, a global order emerges: the oscillators begin to move in a coordinated manner, sharing a common phase.

This transition does not depend on the details of individual elements, but on the collective properties of the system. This is why the same phenomenon of synchronization appears in very different contexts: the language of phase transitions allows us to describe them all with a single fundamental idea—that of order arising from interaction.

In essence, when a set of oscillators synchronizes, we are not witnessing a miracle, but a phase transition. The Japanese mathematician Yoshiki Kuramoto demonstrated this in the 1970s with a simple and elegant model: beyond a certain coupling threshold, disorder spontaneously transforms into collective order. Coherence, here, is not imposed from the outside: it emerges.

To say that coherence is "beyond metaphor" therefore means freeing it from two opposing misconceptions. The first is naive reductionism, which dismisses it as vague language. The second is mystical enthusiasm, which turns it into a catch-all term. In reality, coherence is a precise physical condition that describes how a system maintains stable relationships among its parts over time, resisting noise, fluctuations, and the natural tendency toward disorder.

It is no coincidence that one of the central problems of contemporary physics is understanding how systems maintain coherence and for how long. In laboratories in Zurich or Santa Barbara, where quantum computers are being built, the daily challenge is to protect coherent states from environmental decoherence. Every vibration, every thermal disturbance is an enemy. This language—protection, maintenance, resistance—will become familiar when we begin to discuss living systems. But here it suffices to grasp the essential point: coherence is not a conceptual embellishment; it is a physical struggle against dispersion.

In this sense, coherence offers biology something that biochemistry alone struggles to provide: a vocabulary for discussing dynamic unity without falling back on vitalism. It does not introduce mysterious forces, but rather describes modes of organization. As physicist Philip Anderson wrote in his famous article *"More is Different,"* *"at every level of complexity, new laws emerge."* Coherence is one of these emergent laws, and to ignore it is to forgo understanding how the many become one.

When Order Becomes Physics: A Cultural History of a Key Term.

Every major scientific discipline has certain guiding words—terms that seem simple but, upon closer inspection, carry with them a complex history, made up of shifts in meaning, misunderstandings, and discoveries. Coherence is one of these. Even before it became a term in physics, it was an intuitive idea: that which "holds things together." But the way in which this "holding together" has gradually been refined says a great deal about the profound shift through which science has learned to conceive of order in the natural world.

In the nineteenth century, coherence entered the scientific lexicon through optics. This was the era when light definitively ceased to be merely a mysterious stream and became a measurable wave phenomenon. In Cambridge, in the 1830s, George Biddell Airy studied interference fringes; shortly thereafter, Augustin-Jean Fresnel in Paris demonstrated that the light patterns observed in the experiments were not illusions, but the result of waves superimposing in an orderly manner. Here, coherence denotes a concrete property: the ability of two waves to produce stable interference. If the waves "fall out of step," the effect vanishes.

In this initial sense, coherence has nothing philosophical about it. It is almost artisanal: it concerns light sources, slits, and mirrors. Thomas Young had already intuited this as early as 1801 with his famous double-slit experiment, presented to the *Royal Society* in London. But it was not until the second half of the century that the word "coherence" began to appear systematically in treatises on optics as a technical term. Coherent light is capable of producing orderly patterns; incoherent light, such as that from a candle or the sun, "mixes" too many independent emissions.

The conceptual leap occurred in the 20th century, when coherence ceased to be a marginal property of light waves and became a general interpretive key. With the birth of quantum mechanics, between the 1920s and 1930s, the notion of "phase"—central to defining coherence—became central to the description of matter. Erwin Schrödinger, in his 1926 works, spoke of wave functions that can add up or cancel each other out, just like Fresnel's light waves. This is not a metaphor: it is a conceptual continuity.

It was during this period that the term "coherence" took on a new status. It no longer referred merely to an experimental requirement but to an ontological condition of the system. A coherent quantum system is one that preserves phase relationships among its parts. When these relationships are lost, the phenomenon is called decoherence—a term that would not become firmly established in the scientific vocabulary until decades later, in the 1970s and 1980s, thanks to the work of H. Dieter Zeh and Wojciech Zurek.

An anecdote often recounted in the halls of the University of Texas, where Zurek worked for many years, concerns a lecture in which he explained decoherence using the image of an orchestra that loses its conductor. Someone objected that it was too "humanistic" a metaphor. Zurek replied curtly: *"It's not a metaphor; it's a dynamic.*" Once again, coherence resists attempts to reduce it to figurative language.

The decisive point, from a semantic standpoint, is this: coherence ceases to describe similarities between elements and begins to describe stable relationships. It is not important that the parts be identical, but that they be persistently correlated. This explains many current misunderstandings, especially when the term is applied to biology or the life sciences. Speaking of coherence does not mean evoking harmony in an aesthetic sense, nor order imposed from above. It means indicating a physical condition in which the parts of a system share information in a non-random way.

An interesting cultural shift took place in the 1950s, with the invention of the maser and then the laser. In New York, in 1954, Charles Townes presented the first experimental results on coherent microwave emissions. In documents from that era, the word *"coherent"* appears with unprecedented frequency. It is as if physics suddenly discovered that this property—studied for decades in an almost marginal way—is actually capable of producing radically new technologies. From that moment on, coherence was no longer just a description: it became a resource.

This historical journey helps explain why the word is often misunderstood today. On the one hand, there are those who use it vaguely, as a synonym for "order" or "balance." On the other hand, there are those who reject it as if it were a poetic intrusion into a language that should remain dry. Both positions ignore the concrete history of the term. Coherence originated as an operational concept, refined itself as a theoretical concept, and has always maintained an experimental foundation.

Reexamining this semantic history is not a scholarly exercise. It serves to remind us that when we speak of coherence today—

even outside of physics—we are dealing with a word that, over the course of two centuries, has learned to describe how unity emerges from multiplicity without the need for metaphysical shortcuts. It is a word that has passed through laboratories, conferences, and theoretical debates, and for this very reason deserves to be used with precision.

To measure is to intervene: when observation disrupts coherence.

There is a precise moment in the history of twentieth-century physics when it becomes clear that measuring does not simply mean "looking more closely." It means changing the state of affairs. This conceptual shift, which originated in physics laboratories, has profound consequences for the way we conceive of life as a coherent phenomenon. Without bringing consciousness or the human observer into the picture, it suffices to remain faithful to one essential fact: every measurement is a physical interaction, and every interaction leaves a trace.

In the 1920s, in Copenhagen, Niels Bohr insisted on a point that at the time seemed almost provocative: there are no physical properties independent of the experimental context. An electron does not "have" a position or velocity in the classical sense; these quantities emerge at the moment of measurement. Werner Heisenberg formalized this insight in 1927 with the uncertainty principle, but the question of coherence remained in the background, like an unresolved tension.

For decades, the "collapse of the wave function" was discussed as an enigmatic, almost theatrical event. Then, between the 1970s and 1980s, the scene changed. At research centers in Heidelberg and Los Alamos, physicists such as H. Dieter Zeh and, later, Wojciech Zurek showed that no mysterious "jump" is needed: all that is required is th tion with the environment. A system loses coherence when it enters into uncontrolled interaction with its surroundings. Measurement is a (very refined) special case of this interaction.

An anecdote told by Zeh concerns an informal discussion that took place in 1970 during a coffee break at the University of Heidelberg. A colleague asked him why he placed so much emphasis on the environment—something that no one "sees" directly. Zeh replied by pointing to the room: *"We don't see the air, but try taking it away."* Decoherence works the same way: it is invisible, yet omnipresent.

Here, the word "observation" takes on a new meaning. It no longer refers to a cognitive act, but to a physical process of amplification: a microscopic correlation is imprinted onto a macroscopic system, leaving a stable trace. In the neutron physics laboratories in Grenoble or in the trapped-ion experiments in Innsbruck, the loss of coherence is not a philosophical speculation but a concrete technical problem. Isolating a system means protecting it from the environment; measuring it inevitably means exposing it.

This clarifies a widespread misunderstanding: coherence is not "destroyed" because someone is observing it, but because something is interacting with it. Even an automatic detector, a photographic plate, or a cryogenic sensor plays the same role. In 1996, Serge Haroche and his team at the *École Normale* Supérieure in Paris experimentally demonstrated how coherent quantum states can be tracked step by step as they undergo decoherence. There is nothing sudden about it: coherence gradually unravels, like a choir losing its synchrony.

The conceptual implications are significant. Measurement is not an external addition to the theory, but an intrinsic aspect of the system's dynamics. In this sense, coherence appears as a fragile condition, requiring isolation, protection, and maintenance. The loss of coherence is not a failure of the theory, but its completion: it explains why the macroscopic world appears stable, localized, and "classical."

This point has an interesting cultural resonance. In the very same years that physics was clarifying the role of measurement, philosophers of science such as Bas van Fraassen and Ian Hacking were reflecting on the interventionist nature of the experiment. Hacking, in particular, likes to repeat that *"if you can spray it, then it's real,"* referring to electrons in electron microscopes. In short, reality manifests itself through controlled interactions, not through disembodied gazes.

Talking about coherence and measurement, therefore, does not mean venturing into nebulous territory. It means recognizing that physical order is not fixed once and for all, but depends on the conditions of interaction. Coherence does not disappear because the world is unstable, but because the world is relational. And it is precisely this relationality, laid bare by the physics of measurement, that makes the concept of coherence a powerful tool even beyond the boundaries of physics itself.

The Test of Time: When Stability Becomes the True Sign of Life.

When we think of coherence, our imagination immediately turns to space: aligned elements, parts that "get along," orderly structures. It's a natural reaction, but a misleading one. Coherence, in physics, does not primarily play out in space. It plays out in time. It is a quality of duration, not of form. And that is precisely why it is less intuitive and, at the same time, more decisive.

A system is coherent not because it appears orderly at a single moment, but because it maintains precise relationships among its parts as time passes. Coherence is an active memory: the ability to remain true to oneself through change. In this sense, a photograph is not enough to describe it. A film is needed.

A classic—yet often misunderstood—example comes from the physics of atomic clocks. In 1967, at *the "Bureau International des Poids et* Mesures" in Sèvres, near Paris, the second was redefined based on the transition frequency of the cesium-133 atom. That frequency was not chosen because it is "beautiful" or simple, but because it is extremely stable over time. Here, coherence does not concern the energy itself, but rather its temporal repeatability. If that oscillation were to lose phase after just a few cycles, the clock would cease to be a universal standard.

The same logic applies, on different scales, to all coherent phenomena. A laser, for example, is not a laser because it emits intense light, but because it maintains a constant phase relationship over time. If this relationship fluctuates too rapidly, the light remains, but the coherence disappears. It is a subtle yet crucial distinction: energy can exist without coherence; life, as we shall see, cannot.

In the 1990s, in the laboratories of Santa Barbara, physicist Anthony Leggett insisted on a point that at the time seemed almost controversial: the real challenge of complex systems is not to create ordered states, but to make them last. Nature is full of ephemeral configurations; what is rare is dynamic stability. Coherence, in this sense, is not a state, but a process that spans time without dissolving.

This perspective radically changes the way we think about physical systems. A coherent system is not static: it is reliable. It does not eliminate fluctuations, but absorbs them without losing its identity. This is a theme that recurs even outside of physics. The composer György Ligeti explained in a 1983 interview that music "holds together" not when it repeats a pattern, but when it manages to remain recognizable even as everything changes. This definition is surprisingly close to the physical definition of temporal coherence.

From a theoretical standpoint, the question of time challenges a naive view of natural order. Static order is fragile; a single impact is enough to destroy it. Temporal coherence, or " ," on the other hand, implies a capacity for continuous renewal. Physicists speak of dynamic stability, an expression that sounds like an oxymoron but isn't. It means that the system remains the same precisely because it never stops moving.

An anecdote recounted by Ilya Prigogine during a conference in Brussels in the early 2000s illustrates this point well. Pointing to a vortex in a fluid, he observed: *"This is an order that exists only as long as it flows*." If the flow stops, the vortex disappears. Coherence is not opposed to time; it is a particular way of inhabiting it.

This is why temporal stability is so crucial. It is not enough for the parts to be interconnected; they must be so in the right way and for as long as necessary. Coherence is an operational fidelity: the system "remembers" its own relationships and renews them moment by moment. Without this dynamic memory, every structure dissolves into noise.

In this light, coherence ceases to be a theoretical luxury and becomes a physical necessity. It is what allows a phenomenon to exist as a recognizable entity over time, rather than as a mere momentary coincidence. And it is precisely this ability to traverse time without losing itself that makes coherence a fundamental conceptual key to understanding vital phenomena, without even having to speak of cells, genes, or metabolism.

Coherence, synchrony, correlation, organization: four words, four different worlds.

In scientific language—and even more so in popular science—the word "coherence" is often used as an elegant synonym for order, synchrony, or good organization. It's an understandable temptation: all these concepts evoke the idea of a world that "works," in which the parts do not hinder one another. Yet, from a physical and biological standpoint, confusing these ter s means losing precisely what makes coherence a radical and fruitful concept.

To understand this, it helps to start with a concrete example. In 2009, during a public drill in London, thousands of commuters moved simultaneously along the escalators at the

Canary Wharf subway station. The flows were synchronized, the directions were clear, and the organization was impeccable. Yet no one would say that this system is coherent in a physical sense. It is orderly, yes. Synchronized, to some extent. But all it takes is a power failure for everything to come to a halt. Coherence, on the other hand, is what allows a system to maintain its functional identity even when conditions change.

Synchrony is perhaps the most intuitive of these concepts. Synchronized events occur at the same time or with a common periodicity. Neurons can fire in sync, as demonstrated by Wolf Singer's experiments at the Max Planck Institute in the 1990s; fireflies can flash in unison along the Chao Phraya River in Thailand, a natural spectacle documented as early as the 1950s. But synchrony, in and of itself, is fragile. It can arise from external couplings, from constraints, or from simple dynamic coincidences. It says nothing about the system's internal stability, nor about its ability to withstand perturbations.

Correlation is even weaker. Two variables can be strongly correlated without having any deep structural link. In the 1970s, a famous example used in statistics courses showed a surprising correlation between margarine consumption in the United States and the divorce rate in Maine. Correlation measures an observable relationship, not a real unity. In physics—and even more so in biology—mistaking correlation for coherence is like mistaking a shadow for the object itself.

Finally, "organization" is perhaps the most ambiguous term. An organized system features an orderly arrangement of its parts: hierarchies, modules, distinct functions. A Fordist factory of the 1930s was highly organized; the same is true of a beehive or an army. But organization can be rigid, imposed from the outside, and incapable of adapting. History is full of perfectly organized systems that collapse at the first unexpected change. Coherence, on the other hand, is not imposed: it emerges from a network of internal relationships that support one another.

What distinguishes coherence from all these concepts is the fact that it implies a distributed causal unity. In a coherent system, the parts are not simply aligned in time, correlated in data, or arranged in an orderly fashion: they are linked by relationships that ensure the behavior of the whole cannot be reduced to the sum of local behaviors. It is a subtle but decisive difference.

An anecdote recounted by theoretical biologist Stuart Kauffman during a seminar in Santa Fe in the early 2000s illustrates the point well. Showing two computer simulations—one of a highly organized system, the other of a less orderly but more interconnected system—he observed that only the latter continued to function when random perturbations were introduced. He said: *"Organization survives as long as the world remains as expected. Coherence survives because the world changes."*

From this perspective, coherence is not a property visible at a glance. It is not the same as aesthetic harmony, nor is it the same as regularity. It is a profound relational quality, manifested in a system's ability to remain one while its parts change. And this is precisely why, when speaking of life, coherence cannot be replaced by any of its "conceptual relatives" without losing sight of the heart of the matter.

Confusing these terms is not merely a linguistic error; it is a theoretical error. It means reducing life to a phenomenon of apparent order, when in fact it is an event of dynamic integration. Coherence is not the zero degree of chaos, but an active form of unity that no single synchrony, correlation, or organization can explain on its own.

Because to live means holding together what tends to fall apart.

There is one point—often overlooked—that clearly separates life from everything that is not alive: life does not possess coherence; it produces it. And above all, it defends it. In a universe that, in the words of Arthur Eddington, *"spontaneously tends toward disorder,"* every living organism appears as a small, temporary anomaly—a parenthesis of stability opened against the current. But this stability is never guaranteed once and for all: it is the result of continuous, silent, unceasing work. To live, in the deepest sense, means to resist decoherence.

Decoherence is not merely a concept of quantum physics confined to cryogenic laboratories. It is a universal tendency: relationships loosen, phases decouple, systems lose their unity. A rock does not have to do anything to be what it is. An organism does. Every cell, every tissue, every body expends energy not so much to function as to avoid disintegrating as a meaningful unit.

In 1944, in Dublin, Erwin Schrödinger gave a series of lectures. There he introduced an image that has remained famous: that of the organism feeding on *"negative entropy."* Beyond the simplifications that followed, the insight was powerful: life does not escape the laws of physics, but actively inhabits them, locally resisting the drift toward the loss of coherence. It does not violate the second law of thermodynamics; it circumvents it by paying a price.

That price is metabolism, of course, but not only that. It is also regulation, repair, fine-tuning, and continu l selection of what must remain connected and what can be let go. The British physiologist Denis Noble recounted a seemingly trivial example during a lecture at Oxford in the early 2000s: the human heart, observed at the molecular level, is an inherently unstable system. Yet it beats regularly for decades. Not because it is "well-built," but because it is continually corrected, modulated, and maintained within a regime of functional coherence.

Life, therefore, is not a state of equilibrium, but a struggle for equilibrium. Every organism is an open system that must constantly compensate for internal and external fluctuations. Just look at a feverish body: a rising temperature is not an immediate sign of chaos, but an attempt—sometimes successful, sometimes not—to restore compromised coherence. Here, too, coherence does not mean stillness, but the ability to weather disruption without losing one's identity.

Resisting decoherence also means selecting information. Every organism filters, interprets, and discards. Not everything that happens becomes meaningful. In this sense, living is a fundamental cognitive act: maintaining a form of meaning against the noise of the world. Francisco Varela, as mentioned earlier, spoke of *enaction*: the living being does not represent a world that is already given, but brings forth a habitable one. This emergence, however, is fragile. It requires constant work of integration.

This logic holds true even in the most extreme environments. In the early 1980s, near the hydrothermal vents of the Galápagos, biologists discovered communities of organisms thriving without sunlight. These were not examples of "simpler" life, but of life that had found other ways to maintain its coherence under conditions bordering on chemical disintegration. Once again, the lesson was clear: life is not what resists best, but what works hardest to resist.

Viewed in this light, biological coherence loses all its mystical aura and takes on an almost dramatic concreteness. It is a process, not a state. A continuous struggle against the natural tendency of things to separate. Life endures as long as this struggle succeeds. When it no longer succeeds, decoherence takes over. And what remains is matter, no longer an organism.

Paying the price of unity: the energy cost of biological coherence.

There is a fact that always surprises those encountering it for the first time: most of the energy consumed by a living organism is not used for movement, growth, or the production of visible work. It is used to keep the organism from falling apart—to maintain internal integration and keep processes aligned that, left to their own devices, would quickly diverge. This is the energy cost of coherence, a continuous expenditure that rarely features in the popular narrative of biology but constitutes its very backbone.

Physiologists call it basal metabolism. The term is technical, almost understated, but the reality it describes is striking: in a resting adult human, in a temperate room, about 60–70% of daily energy is consumed without anything apparently "active" happening. The heart beats, of course. The lungs breathe. But above all, billions of cells are working to maintain ionic gradients, repair molecular structures, correct errors, and synchronize signals. It is energy expended to prevent internal decoherence.

A prime example comes from the functioning of nerve cells. In the 1950s, at the *Marine Biological* Laboratory in Woods Hole, Massachusetts, Alan Hodgkin and Andrew Huxley studied the action potential in the neurons of the giant squid. What emerged—and what remains striking to this day—is that a huge portion of the brain's energy is used not for "thinking," but for restoring initial conditions after every electrical impulse. The famous sodium-potassium pump, which expels three sodium ions and allows two potassium ions to enter, ceaselessly consumes ATP. Without this behind-the-scenes work, neuronal excitability would collapse within minutes. Thought would, literally, shut down due to a lack of electrical coherence.

This holds true at every scale. In 1961, in a laboratory in Brussels, biochemist Christian de Duve observed how cells invest energy in internal compartmentalization: membranes, vesicles, lysosomes. This is not an architectural luxury, but a functional necessity. Separating, maintaining boundaries, preventing everything from reacting with everything else. Every biological boundary is an energy expense, but also a gain in coherence.

Even when we sleep, this work does not stop. In fact, in some respects, it intensifies. Indirect calorimetry studies conducted at Stanford in the 1990s showed that during deep sleep, certain cellular maintenance activities increase. Sleeping is not "shutting down," but rather reducing external interference in order to invest energy in internal reorganization. An apparent pause for an essential task: resetting the system.

An anecdote told by an Italian researcher who worked in a cell biology laboratory in Heidelberg in the early 2000s illustrates this point well. While studying cell cultures that were kept alive but deprived of stimuli, he noticed that their energy consumption remained surprisingly high. *"They seemed to be at a standstill,"* he said, *"but they were spending everything just to avoid becoming something else."* It's a casual remark, but it hits the nail on the head: life consumes energy to remain exactly as it is, not to transform itself.

This continuous investment also explains why coherence is fragile. A significant reduction in energy intake is enough for integration to begin to falter. During famine or under extreme conditions, the body does not arbitrarily "choose" what to shut down: it sacrifices complex functions to preserve a minimum of unity. It is an emergency strategy against fragmentation.

From a cultural perspective, this idea overturns a deeply rooted metaphor. We are accustomed to thinking of life as accumulation, as growth, as expansion. But biologically, life is above all maintenance. Like a suspension bridge that requires constant maintenance to remain standing, the living organism invests energy to maintain connections. Failing to do so would mean letting time, chemistry, and thermal noise get the better of it.

In this sense, the energy cost of coherence is not a quantitative detail but an ontological clue. It tells us that life is not a passive structure but a process that pays, moment by moment, the price of its own unity. As long as it can afford to do so, it remains alive. When it can no longer do so, coherence dissolves, and with it, the organism.

No life is an island: coherence arises from dialogue with the environment.

There is a persistent, almost instinctive misunderstanding in the way we conceive of biological stability: the idea that an organism is all the more stable the more it manages to isolate itself from the environment. As if life were a fortress, and health depended on the strength of its walls. Yet, upon closer observation of living systems, the opposite is true: coherence does not arise from isolation, but from a continuous, delicate, and often risky relationship with its surroundings.

A living organism is, by definition, an open system. It exchanges matter, energy, and signals. But above all, it exchanges information. Its stability does not lie in defending itself from the world, but in knowing how to interpret it. It is a constant negotiation, made up of adjustments, anticipations, and compromises. When this dialogue is interrupted—whether through excess or deficiency—coherence falters.

A classic example comes from ecology, and it feels like a lesson learned too late. In the 1950s, in the United States, the massive use of pesticides such as DDT seemed like a victory of engineering over the environment. Then it was realized that species considered "protected" were collapsing. They weren't weak; they were out of sync. Their biological coherence depended on subtle environmental signals—cycles, presences, absences—that had suddenly been erased. Rachel Carson, in 1962, described this lucidly in **Silent Spring**: the silence was not peace, but a loss of connection.

The same logic applies on a more intimate scale. Consider the immune system. For a long time, it was described as an army tasked with rigidly distinguishing the "self" from the "non-self." Today we know that this metaphor falls short. Studies conducted in Paris and San Diego in the 1990s showed that the immune system matures through controlled exposure to the environment. Children raised in overly sterilized environments are more prone to immune dysfunction. Once again, coherence does not arise from isolation, but from relational learning.

An anecdote told by a marine biologist who worked at a research station in the Lofoten Islands, Norway, illustrates this point well. While studying plankton populations, he noticed that the most stable ones did not live in "constant" waters, but in areas subject to variable currents. "These are systems trained to adapt to change," he said. Their coherence was not rigidity, but the ability to remain recognizable while navigating fluctuations.

This insight has a long cultural history. As early as 1928, Harvard physiologist Walter Bradford Cannon spoke of homeostasis not as immobility, but as a dynamic equilibrium. Today, we can interpret this concept in an information-theoretic light: an organism remains coherent because it continuously integrates environmental signals into its internal functioning. If it stops doing so, it loses touch with the context that makes it possible.

Human behavior also offers telling examples. During missions involving prolonged isolation—from the Antarctic " " bases of the 1970s to confinement experiments such as Mars-500, which concluded in 2011 near Moscow—it became clear that the loss of rich environmental stimuli compromises psychophysiological coherence. The main problem is not the cold or fatigue, but the relational poverty of the context. Without a "responsive" environment, the human organism struggles to maintain a unified sense of self.

From this perspective, biological coherence appears to be an ecological property even before it is an internal one. An organism is stable because it is situated, because it belongs to a network of relationships that continually inform it. Its identity is not an immutable core, but a form that persists over time precisely because it engages in dialogue with its surroundings.

Perhaps this is why life appears to us as both so resilient and, at the same time, so vulnerable. Resilient, because it knows how to adapt to different contexts. Vulnerable, because no coherence can survive for long in a world that no longer speaks its language. Field biology thus suggests a counterintuitive yet profound conclusion: to live does not mean to defend oneself against the environment, but to remain in relationship with it without losing oneself.

When Unity Breaks Down: Death as the Decoherence of the Living.

Discussing death in a chapter dedicated to the coherence of life may seem like a melancholic digression. In reality, it is an almost inevitable step. If life is the active maintenance of coherence, then death—physical, biological death—can be soberly described as the moment when this coherence is irreversibly lost. Not a sudden, isolated event, but a process of global decoherence involving the entire organism as an informational unit.

From a physical standpoint, a living organism does not die because it "runs out of energy." Energy continues to be present even in a lifeless body. What is lost is the ability to coordinate flows, timings, and relationships. The parts remain, but they cease to speak the same language. It is a dissolution of unity, not a disappearance of matter.

A striking illustration comes from clinical medicine. In 1968, at *Harvard Medical School*, a committee led by Henry Beecher proposed, for the first time, clear criteria for defining brain death. This was not a philosophical choice, but a practical necessity: to distinguish an organism that was still unified from one that had irreversibly lost its functional integration. The heart might still beat, the lungs might still be ventilated, but the central information unit—the capacity to coordinate the whole—was definitively compromised. Death, from this perspective, does not coincide with the cessation of an isolated function, but with the loss of coherence of the whole.

Those who have witnessed a patient's final moments in a hospice often describe a curious detail: it is not silence, but a sort of progressive discord. Parameters that until recently supported one another begin to diverge. The respiratory rhythm no longer "syncs" with the heartbeat, thermal regulation becomes erratic, and biochemical signals lose their synchronization. It is as if the orchestra were still playing, but without a conductor and without a shared score. The instruments are still there, but the music is not.

There is nothing metaphysical about this description, and that is precisely why it is so powerful. Global decoherence is not an evocative concept: it is observable, measurable, and clinically recognizable. Studies conducted in the 1980s and 1990s on brain electrical activity show that, at the moment of death, we do not simply witness a "decline" in the signal, but rather a fragmentation of coherent patterns. The EEG does not immediately flatten out: it first loses structure, then integration. Form dissolves before intensity.

At this point, it is important to clearly distinguish the scientific realm from the metaphysical one. Describing biological death as informational decoherence says nothing— ly in one sense or the other—about what may happen at the level of consciousness, meaning, or transcendence. Speaking of a "phase transition" in a metaphysical sense may have symbolic or philosophical value, but it belongs to a different realm. Field biology, if it is to remain scientifically rigorous, stops one step short: it observes that the organism, as an embodied unit of information, ceases to exist when it is no longer able to maintain its global coherence.

Curiously, this view brings death back into the continuum of life, without either mythologizing or trivializing it. Dying is not the opposite of living, but the point at which the process of maintaining coherence fails definitively. Ilya Prigogine, speaking of systems far from equilibrium, observed that living order is always temporary, sustained at the cost of continuous expenditure. When the cost exceeds the system's capacity, order collapses. There is no cosmic tragedy in this, nor is it merely a mechanical failure.

An anecdote told by a nurse who worked in an intensive care unit in Lyon in the 2000s is enlightening. He said: *"The moment of death is not when the monitor turns off, but when you realize that you can no longer bring back together what has been separated.*" It is a simple, almost folksy phrase, but it hits the nail on the head: death as the irreversible loss of the possibility of restoring unity.

In this light, physical death appears as a fully natural and, at the same time, radical event. It is the end of a unique, unrepeatable form of existence that existed only as long as it was able to hold its internal and external relationships together. Everything else—questions about meaning, continuity, and the possible "beyond"—remains open, but belongs to another chapter of our experience and our thought.

More than the sum, less than the myth: coherence as a scientific fact.

Whenever the word "emergence" is mentioned in biology, someone turns up their nose. All too often in the past, it has been used as an elegant shortcut to mask what could not be explained. But the emergence we are discussing here—coherence as a fundamental property of living things—is neither vague nor esoteric. It is a real emergence, in the most concrete and scientifically respectable sense of the term.

To understand this, we need only take a step back—not toward metaphysics, but toward the recent history of science. In 1977, when Ilya Prigogine received the Nobel Prize for his studies on dissipative structures, he did not propose any conceptual magic. He demonstrated, with data in hand, that in open systems far from equilibrium, forms of global order can emerge spontaneously that are not deducible from individual local interactions. Bénard cells—those elegant convective structures observable in a simple layer of heated fluid—are not "contained" within the water molecules. Yet they are real, measurable, and reproducible. No one today would call them an illusion.

Biological coherence belongs to this same family of phenomena. It is not an additional force, nor is it a hidden entity acting behind the scenes. It is a collective property that arises when a vast number of physical and chemical processes organize themselves in a coordinated manner over time. The crucial point is that this organization is not arbitrary: it has constraints, regularities, and s critical thresholds. In other words, it is a legitimate subject of scientific inquiry.

A lesser-known but illuminating example comes from the work of Walter Freeman, a neurophysiologist at the University of Berkeley, in the 1970s and 1980s. While studying the electrical activity of the olfactory bulb in rabbits, Freeman observed that the overall patterns of neuronal activity could not be predicted by analyzing the behavior of individual neurons. The information did not lie in the "parts," but in the coherent configuration of the whole. Freeman spoke of emergent patterns, not to evoke a mystery, but to describe a stubborn experimental fact: the brain functioned as a dynamic unit, not as a collection of independent components.

Herein lies the core of coherence's scientific legitimacy: it produces observable causal effects. A coherent organism reacts to the environment in a unified manner, maintains its form, and makes biological decisions—from metabolism to reproduction—that cannot be attributed to a single molecule or an isolated biochemical pathway. Coherence makes things happen. And whatever has causal effects falls, by definition, within the realm of science.

It is no coincidence that analogous concepts have emerged in fields far removed from biology. In 1985, at the Santa Fe Institute, physicists, economists, and biologists began to seriously discuss complex adaptive systems. The underlying idea was simple yet radical: some systems cannot be understood by breaking them down, because their most important properties arise from interactions. Life, in this sense, is not an awkward exception, but one of the most sophisticated examples of a general principle.

It is worth emphasizing: recognizing the emergence of coherence does not mean giving up on explanation. It means accepting that the level of explanation changes. As theoretical biologist Robert Rosen wrote in the 1990s, *"a living system is explainable, but not decomposable without loss."* There is nothing unscientific about this. It is, if anything, a call for greater conceptual precision.

A small anecdote, recounted by an Italian researcher who worked in Heidelberg in the early 2000s, captures the cultural atmosphere of the lab well. During a meeting, a young Ph.D. student asked, *"But this emergent property... where exactly is it?"* The group leader replied, after a pause: *"It's where the melody is in a string quartet. Not in a violin, but not outside the instruments either."* The remark brought a smile, but it ended the discussion. Everyone understood.

Coherence, therefore, is not a fancy name for ignorance. It is the recognition that life operates at a level of organization where the whole takes precedence over the list of parts. Accepting this does not mean abandoning rigor, but extending it. It means taking seriously what biological experience has always shown: that the living is a whole that acts as a whole, and that this wholeness is not a mystery, but a phenomenon.

Coherence as seen through mathematical models.

If biological coherence were merely a linguistic suggestion, it would hardly have attracted the attention of mathematicians, theoretical physicists, and scholars of complex systems. Instead, for at least half a century, it has been mathematics itself—the most rigorous and least prone to metaphors—that has provided one of the most solid foundations for the idea that the order of living systems emerges as a global, stable, and formally describable property.

An often-overlooked turning point dates back to the 1960s at MIT in Cambridge, when Norbert Wiener and his students laid the foundations of modern cybernetics. The idea that complex systems could maintain coherent behavior through *feedback* loops was not a philosophical intuition, but a mathematical result. The differential equations describing these systems showed how the system as a whole could stabilize into dynamic configurations—attractors—regardless of local fluctuations. In other words: coherence was not imposed from the outside, but generated by internal organization.

In the 1970s, this mathematical language found an ideal home in the work of René Thom, a French mathematician and founder of catastrophe theory. Thom was not directly concerned with molecular biology, but with forms. He showed how continuous systems could undergo sudden qualitative changes while still following deterministic laws. His famous 1972 lecture at *the* "*Institut des Hautes Études* Scientifiques" in Bures-sur-Yvette sparked months of debate: the idea that the form of an organism was a stable effect of a field of mathematical constraints—and not merely a product of local instructions—proved surprisingly compatible with what embryologists observed under the microscope.

Closer to our own time, dynamic networks have provided an even more compelling picture. When Stuart Kauffman began studying *"random Boolean* networks" in the late 1980s, he was seeking a simple answer to a huge question: how does a cell maintain its genetic identity despite molecular noise? His models showed that, beyond a certain threshold of connections, networks spontaneously tend toward ordered yet flexible states, which Kauffman called "the edge of chaos." It was not a poetic metaphor: it was a mathematical result. Coherence emerged as a global state of the system, not as a property of a single gene.

An anecdote recounted by Kauffman himself during a seminar in Santa Fe in 1994 captures the atmosphere of that research well. After showing a simulation in which a complex network stabilized into a few robust attractors, a biologist in the audience commented, *"It almost seems as if the network knows what to do*." Kauffman smiled and replied, *"No, it seems as if the whole knows*." It was a joke, but it hit the nail on the head: mathematics was providing a language to discuss the implicit knowledge of living systems without resorting to mysterious agents.

The theory of nonlinear dynamical systems has also made a decisive contribution. Strange attractors, studied by Edward Lorenz as early as 1963 in the context of climate models, showed that order and unpredictability can coexist. In biology, this means that an organism can be highly coherent while remaining sensitive to its context. The equations do not describe a rigid machine, but a constrained dance: a dynamic coherence, not a static one.

What makes these models particularly significant is the fact that coherence becomes measurable, simulable, and falsifiable. It is no longer an evocative concept, but a mathematical property: structural stability, convergence toward attractors, synchronization of networks. When, in the early 2000s, research groups in Barcelona and Boston began modeling real metabolic networks, they discovered that the loss of global coherence often preceded the functional collapse of the system. Once again, mathematics preceded biological interpretation.

In this sense, mathematical models offer a grammar for describing how unity emerges from multiplicity without being reduced to it. Biological coherence, far from being a vague concept, thus appears as a well-defined formal structure: a collective event that equations, surprisingly, are able to describe.

When Mathematics and Biology Learn to Understand Each Other.

For a long time, mathematicians and biologists viewed one another with a certain degree of mistrust, as practitioners of neighboring yet incompatible fields. The former spoke of abstract spaces, variables, and structural stability; the latter of cells, tissues, development, and diseases. When they met, they often failed to understand one another. Not for lack of intelligence or openness, but for a deeper reason: language.

An anecdote recounted by a Dutch theoretical biologist during a workshop in Leiden in the early 1990s is emblematic. At the end of a presentation on dynamic models of embryonic development, an experimental biologist approached him and said, *"Your equations are elegant, but where is the organism?"* The question was not meant to be confrontational; it was sincere. For her, the organism was something that could be dissected, stained, and observed under a microscope. For the mathematician, however, the organism was precisely that structure of relationships that the equation sought to capture.

This semantic gap has long hindered the understanding of biological coherence. In biology, the word "coherence" evoked, at most, good functional integration; in mathematics and systems physics, it denoted a rigorous global property, definable in terms of stability, correlation, and collective behavior. Talking about coherence without clarifying the context of the discussion often meant talking past one another.

Yet, starting in the 2000s, something changed. Places like the *Santa Fe Institute* in New Mexico, *the EMBL* in Heidelberg, and the *Max Planck Institute for the Physics of Complex Systems* in Dresden have become spaces for linguistic cross-pollination. Here, biologists, mathematicians, and physicists began working not so much to translate their own concepts as to build a shared vocabulary. No longer was it a matter of "applying mathematics to biology," but rather of co-constructing models that arose from real biological questions.

An often-cited example is the work on gene regulatory networks in development. When, in the early 2000s, interdisciplinary groups began describing these processes as high-dimensional dynamic systems, many biologists were skeptical. Then the results came in: the mathematical models were able to predict the robustness of certain developmental patterns better than isolated molecular descriptions could. The organism's coherence did not emerge from individual genes, but from the overall structure of the interactions.

An Italian mathematician recalled in a 2011 interview how the real conceptual leap had taken place not in front of a blackboard, but during an informal dinner in Trastevere after a conference in Rome. He recounted that at one point an embryologist said, *"Perhaps when you say 'attractor,' you mean what we call 'cellular fate.'"* It was a mutual revelation. The words were different, but the concept was the same.

This dialogue has enriched both fields. Biology has gained formal tools for conceiving of the unity of life without reducing it to a list of components. Mathematics, for its part, has been forced to grapple with historical, contingent systems open to the environment—very different from the ideal objects found in textbooks. In this encounter, biological coherence ceased to be a concept "halfway" between poetry and science and became a shared property, describable from multiple perspectives.

Today, when we speak of an organism as a stabilized informational structure, that dialogue is no longer optional but an epistemological necessity. Coherence does not belong to a single discipline: it is the phenomenon that compels disciplines to engage with one another. And perhaps this is precisely its most profound lesson.

The Shadow of Vitalism: How Emergence Escaped Scientific Suspicion.

At the beginning of the 20th century, uttering the word "emergence" in a scientific context was risky. It sounded like an admission of ignorance, or worse, a disguised return to vitalism. In an era that had just witnessed the triumph of organic chemistry and experimental physiology, the idea that new properties could "appear" at the level of the organism seemed like a conceptual shortcut, an elegant way of saying: we cannot explain it.

This suspicion was not unfounded. Many early emergentists—one need only think of the British philosopher C. D. Broad, who in 1925 spoke of properties that were not deducible even in principle—left a gray area open. If a property is not deducible, is it then outside the realm of physical causality? Is it something added from the outside? During those same years, laboratories such as Jacques Loeb's at *Rockefeller University* in New York were pursuing the opposite agenda: to demonstrate that life was entirely attributable to physicochemical forces. Any reference to "totality" or "organism" was viewed with annoyance.

An anecdote recounted in the memoirs of a young German researcher captures the atmosphere of the time well. During a discussion in Berlin in 1931, a colleague dared to speak of "emergent properties" of cellular metabolism. The silence in the room was followed by a cutting remark: *"We do science here, not metaphysics*." The accusation was clear: emergence equals mystery.

There were three main criticisms. First: emergentism seemed to violate physics' principle of causal closure. Second: it lacked formal models, relying instead on qualitative descriptions. Third: it risked reintroducing an implicit purpose into nature—something modern science had painstakingly expelled. In that context, reductionism appeared not only more rigorous but also more honest.

What has changed is not so much the idea of emergence as the scientific ground on which it rests. The objections of the early 20th century presupposed a world governed by linearity, simple causality, and isolatable systems. But post-World War II science—from the thermodynamics of open systems to the theory of complex systems—has shown that these assumptions were incomplete. Today we know that new behavioral regimes can emerge without violating any fundamental laws, simply when the number and complexity of interactions exceed a certain threshold.

The decisive turning point was formal. When emergence becomes modelable—through nonlinear dynamics, interacting networks, and collective stability—it loses its aura of mystery. Biological coherence, in this framework, is not a magical event but a stable configuration of relationships. Not "something extra," but rather "something that happens" when the system is sufficiently organized.

Even the old accusation of finalism has proven less solid than it seemed. Today, speaking of goal-oriented behaviors or the maintenance of coherence does not imply any external purpose, but rather describes an internal constraint within the system. A coherent organism does not seek to remain so: it simply ceases to exist if it does not. It is a structural necessity, not a design.

In this sense, contemporary emergentism does not call for philosophical leniency, but for scientific attention. Historical criticisms were the product of an era lacking adequate conceptual and mathematical tools. Today, those tools exist. And with them, the idea that life is a coherent emergent property no longer appears as a surrender of explanation, but as its natural extension.

This is what our reductionist training prevents us from recognizing.

There is a process that accompanies nearly every scientific training: breaking things down. An organism is broken down into organs, an organ into tissues, a tissue into cells, and a cell into molecules. It is a powerful process that has produced some of modernity's most impressive achievements. But like all processes repeated without reflection, it risks becoming a cognitive habit. And habit, as we know, does not help us see what does not fit the pattern.

Many scientists describe, with a touch of irony, that same moment of bewilderment. It often occurs during the early years of research: one observes a biological phenomenon that "works" flawlessly—robust embryonic development, coordinated healing, a refined immune response—but cannot explain it by adding up the individual molecular steps. *"I can describe every piece,"* a young cell biologist confided at a seminar in Cambridge in 2014, *"but the overall behavior eludes me."* It wasn't a limitation of expertise; it was a limitation of perspective.

Scientific education in the twentieth century, especially in the life sciences, was shaped by a specific idea of explanation: to understand is to reduce. This approach has deep roots, dating back at least to the success of classical mechanics and analytical chemistry. It is not surprising that it was adopted as a universal model. The problem arises when what was once an effective method becomes the only conceivable method.

A little-known but illuminating anecdote concerns a visit by Ludwig von Bertalanffy to an American university in the 1950s. Invited to give a lecture on systems theory, he was asked by a student, *"In which chapter of the biochemistry textbook does the system fit?"* Bertalanffy replied calmly, *"In none. And that is precisely the problem."* It was not a philosophical provocation, but a pedagogical observation: certain concepts do not fit because there is no room for them in standard education.

Educational reductionism tends to create a paradox. On the one hand, it provides increasingly sophisticated analytical tools; on the other, it casts suspicion on any property that cannot be pinpointed. If one cannot indicate where coherence lies—in which molecule, in which gene—then one is tempted to consider it a descriptive illusion. But emergence cannot be pointed to with a finger: it is recognized by observing the system's behavior over time.

This difficulty does not stem from ignorance, but from a training that prioritizes static analysis over dynamic analysis, the element over the relationship. Recognizing an emergent ph —such as coherence—requires a shift in intellectual posture: not abandoning deconstruction, but pairing it with a capacity for conceptual recomposition. This capacity is rarely taught explicitly.

It is no coincidence that many scientists have "discovered" emergence late in their careers, often outside the classroom. Some encounter it while observing an ecosystem in the field, as marine biologist Rachel Carson recounted in her diaries; others intuit it while working with computational models that reveal unexpected behaviors; still others, more simply, notice that an organism continues to function even when certain parts are disrupted. Coherence manifests itself, but one must be willing to see it.

In this sense, the problem is not reductionism itself, but its exclusivity in education. As long as we train scientists who excel at breaking things down but are afraid to speak of the whole, emergence will continue to appear as something vague. When, on the other hand, we learn to consider it an observable, stable, and reproducible phenomenon—even if it cannot be localized—then coherence ceases to be a suspect concept and becomes what it is: a real property of life.

IV. The Organism as a Non-Local System.

There comes a point in the course of this book where criticism of reductionism can no longer be limited to pointing out what does not work. It must propose a shift in perspective. Chapter IV arises precisely from this need: to introduce the reader to an idea that is as simple as it is destabilizing for classical biology, namely that in living organisms, the behavior of the parts cannot be understood without starting from the state of the whole.

After demonstrating, in the previous chapters, the limits of a purely biochemical explanation and the difficulties Darwinism faces in accounting for form and organization, we now take a further step. It is no longer merely a matter of saying that molecules are not enough, but of recognizing that biological causality is neither linear nor local. In organisms, what happens at one point depends on conditions that cannot be confined to that point. Life, in other words, functions as a global system even before it functions as an assembly of components.

This insight did not originate in biology, but finds surprising parallels with twentieth-century physics. When Albert Einstein spoke, with a certain irony, of *"spukhafte* Fernwirkung"—the famous *"spooky action at a* distance"—he was expressing his unease in the face of a non-locality that seemed to violate classical common sense. Today, without forcing inappropriate analogies, many biologists recognize that living systems also exhibit a form of functional non-locality: not because they transmit instantaneous signals, but because the global organization constrains local behavior.

As theoretical biologist Robert Rosen wrote, *"An organism is not a machine because in a machine the parts explain the whole, whereas in an organism it is the whole that explains the parts.*" This sentence perfectly sums up the purpose of this chapter. In living systems, there is no "director molecule," no sovereign gene, and no chemical reaction capable, on its own, of explaining biological order. Causality is distributed, circular, and often retroactive. Each level of the organism exists in relation to the others.

This idea of the primacy of the organism has a long, hidden history. Kurt Goldstein, a neurologist and thinker of the first half of the twentieth century, observed patients with brain injuries and noted that residual behavior was not simply the sum of intact functions. The entire organism reorganized itself, finding new equilibria. He wrote: *"The injured part can only be understood in relation to the organism as a whole.*" This was not a clinical metaphor, but a theoretical statement.

In this chapter, this perspective is taken to its logical conclusion: the organism is not the end result of local processes, but the very condition that makes those processes possible. Cells, tissues, and even gene expression respond to a global state of coherence or incoherence. As Nobel laureate Ilya Prigogine observes, *"cause and effect are no longer separable in time and space.*" Applied to biology, this statement suggests that life does not follow a linear causal chain, but rather a network of reciprocal constraints.

This chapter therefore plays a strategic role in the book's structure. It serves to lay the conceptual groundwork for the introduction of biological fields, which cannot be understood unless one first accepts the idea that the organism is a primary, non-derived unit. Discussing fields without this preliminary step would reduce them to mere metaphors. Here, however, it is made clear why they are a theoretical necessity.

Locality and Non-Locality: The Differences.

For centuries, science has viewed nature with a reassuring conviction: to understand a phenomenon, one need only identify where it occurs and what causes it. This is the idea of locality, according to which every effect has a cause that can be identified at a precise point in space and time. This approach has worked exceptionally well for mechanics, chemistry, and much of classical physiology. But when we truly delve into the complexity of living organisms, it begins to show deep cracks.

This chapter of the book is set precisely at that point of friction. After challenging the idea that life can be reduced to a chain of biochemical reactions, it addresses a crucial distinction: that between local processes and non-local dynamics. This is not a theoretical subtlety, but a difference that radically changes the way we interpret biological organization.

In a local system, the behavior of one part depends essentially on what is next to it. An enzyme catalyzes a reaction because it encounters a substrate; a receptor is activated because a molecule binds to it. Everything happens "here," and what happens elsewhere is, in principle, irrelevant. This is the implicit model of classical biochemistry—powerful and precise, but limited.

Biological nonlocality, on the other hand, introduces a different logic. In organisms, what happens at one point often depends on the state of the entire system—not because mysterious forces act at a distance, but because living beings are organized as a coherent unit, in which the global context constrains local behavior. An example well-known to experts but rarely discussed outside specialized circles concerns the heart rhythm. Denis Noble, a physiologist at Oxford, has shown that the heartbeat is not "controlled" by a single pacemaker cell, but emerges from the collective interaction of millions of heart cells, each of which is sensitive to the overall state of the tissue. Isolating a single cell means losing sight of its actual behavior.

The difference between local and non-local becomes even more evident in embryonic development. If locality were the dominant principle, knowing the position and fate of every single cell would suffice to explain the final form of an organism. Yet, classic 20th-century experiments have shown that embryos separated at an early stage can reorganize themselves and give rise to complete structures, even if parts are "missing" that, according to a local logic, would have been indispensable. Development responds to a field of global relationships, not to a point-by-point map.

Walter Freeman, an American neuroscientist, used a deliberately provocative expression when speaking of brain activity: *"The brain does not process information like a switchboard, but like a storm.*" What he meant was simple and radical: there is no precise location where a thought or perception "arises." Neural activity takes on global configurations that give meaning to individual signals. Once again, the whole precedes the parts.

Locality seeks the cause; non-locality seeks the condition. In the former, one asks, "Which element produces this effect?" In the latter, "In what state must the system be for this effect to be possible?" It is a subtle but decisive difference. Claude Bernard, as early as the nineteenth century, had intuited something similar when he stated that "the stability of *the milieu intérieur* is the condition of free life." He was not speaking of specific molecules, but of a global state that makes every local function possible.

This distinction also helps us understand why the idea of a "master molecule" is so seductive and, at the same time, so misleading. Attributing the role of ultimate cause to a gene, a hormone, or a neurotransmitter is a local reflex applied to a system that is not local.

From the body to society: the non-locality that governs us.

There is a point, in the comparison between biology and economics, where the parallel ceases to be an elegant metaphor and becomes a tool for understanding. It is the point of non-locality. Just as in a living organism the behavior of a cell depends on the overall state of the body, so too in economic and social systems the actions of individual actors are deeply conditioned by global dynamics that no one directly controls. John Maynard Keynes clearly intuited this in the 1930s, speaking of *"animal spirits*": seemingly rational individual decisions that, taken together, produce unpredictable collective effects. No single investor, taken individually, "decides" to cause a financial crisis, and yet a crisis occurs, like a systemic fever. The example of the 1929 Wall Street crash—still studied today not only by economists but also by complex systems theorists—reveals a dynamic strikingly similar to certain biological phenomena: small local signals become significant only because the system as a whole has lost its coherence. Karl Polanyi, observing Europe between the two world wars, spoke of a "disintegration of the social fabric" even before the economic one. Here too, as in living systems, it is not the individual part that explains the collapse, but the overall state of the network of relationships. Closer to home, sociologist Mark Granovetter has shown how individual behaviors—such as joining a protest or following a fashion trend—depend on invisible collective thresholds: what appears to be a personal choice is in reality a non-local event, made possible by the overall configuration of the social network. In biology as in the social sciences, non-locality does not eliminate the individual, but places them within a context that precedes them and makes them intelligible. Once again, the whole comes before the parts.

When the future determines the present.

If non-locality challenges the idea that every cause lies at a precise point in space, time takes it a step further: it shows that, in living systems, the cause does not even lie in an isolatable moment. An organism never lives solely in the present. Every response is steeped in the past and oriented toward a possible future. The immune system, for example, reacts based on memory: an infection that occurred years ago alters the quality of the cellular response today. In 1973, during a conference at Cold Spring Harbor, Niels Jerne described the immune system as a network that "remembers" and anticipates, more like a conversation than a chain of reactions. Here, time does not flow in a straight line: it circulates, folds back on itself, and returns. The same logic applies to the entire body. An injured athlete recounted that the pain in his knee would return even before he exerted himself, merely at the prospect of the race. The anticipation of the future had already altered his physiological present. Similar phenomena are well known in societies as well. Economic crises often begin before the numbers justify them: it is enough for the idea of an uncertain future to spread for collective behavior to change today. In 1992, on the eve of the lira crisis, some observers noted that the markets reacted more to rumors than to official data. As in an organism under stress, the anticipation of danger reorganizes the system as a whole. Sociologist Robert K. Merton spoke of a "self-fulfilling prophecy": future expectations that become present causes. In both cases—biological and social—time acts in a non-local way: the past is not truly past, the future has not yet happened, and yet both actively participate in shaping the here and now. Temporal non-locality is perhaps the most subtle form of the primacy of the whole: the system does not respond to what happens, but to what it means within the flow of its history.

When even machines become non-local.

In recent decades, observing the functioning of large computer networks, many engineers have had a sensation familiar to biologists: no command center, yet surprisingly coherent global behavior. The Internet, ever since its original architecture was designed in the 1960s by Paul Baran and Donald Davies at *the RAND Corporation* and *the National Physical Laboratory* in Teddington, has been designed to function even if significant parts fail. Data packets find alternative routes, traffic reorganizes itself, and the network "persists." In this sense, the non-locality of computer networks resembles that of living organisms: what happens at a single node depends on the overall state of the system. However, the similarity ends there. A living organism does not merely maintain functionality: it interprets what is happening. When a blackout struck a large data center in Northern Virginia in 2012, redundancy systems kicked in according to predefined protocols. Everything worked, but no network "understood" the significance of the event. In an organism, by contrast, a disturbance is never merely technical: it is meaningful. Canadian biologist Brian Goodwin observed that complex machines can simulate organization, but not the formation of meaning. Artificial networks are non-local in space, but they remain local in meaning: they operate according to external rules, not according to a lived coherence. Even the most advanced artificial intelligence architectures, such as those developed in *DeepMind's* laboratories in London, demonstrate adaptability and learning, but always within a scope of goals assigned from the outside. The organism, on the other hand, is at once a network, a set of rules, and a purpose: its "field" is not programmed; it is continuously regenerated. The comparison with artificial systems is therefore valuable not for reducing life to a sophisticated machine, but for clarifying what is still missing in machines: the primacy of the whole as an internal condition, not as an imposed design.

When the whole was a nuisance to the parts.

To understand why speaking of an organism as a non-local system still seems, at times, to go against the grain today, we must return to the intellectual climate of the second half of the twentieth century. Between the 1950s and 1970s, biology experienced a period of spectacular successes: the discovery of the structure of DNA in 1953, the decoding of the genetic code, and the rise of molecular biology as a unifying language. In that context, the idea that the whole could come before the parts seemed not only superfluous but suspect. Francis Crick, in a lecture at Cambridge in 1966, stated in no uncertain terms that the ultimate goal of biology was to explain life "in terms of physics and chemistry." It was not arrogance, but confidence in a winning paradigm. Those who proposed holistic views—such as the Austrian biologist Ludwig von Bertalanffy with his *"General Systems* Theory"—were often relegated to the margins: interesting from a philosophical standpoint, it was said, but of little practical use in the laboratory. During those same years, in Chicago, neuroembryologist Paul Weiss spoke of morphogenetic fields and of global coordination in development, but his ideas struggled to find a place in leading journals. A recurring anecdote in university corridors tells of a young European researcher who, in the 1970s, was advised to "remove all references to the organism" from an article if he wanted to get it published. The climate was such that holism was conflated with theoretical vagueness, if not with vitalistic tendencies.

It was only later, with the internal crises within reductionism itself—the inability to predict the behavior of complex genetic networks or to explain the robustness of organisms—that perspectives slowly began to shift.

Figures such as Ilya Prigogine demonstrated that order and coherence could emerge without a central director. But it took time for these ideas to filter into the biology of living organisms. Academic resistance was not merely scientific: it was cultural, institutional, and tied to careers, funding, and shared languages. Understanding this history means grasping why field biology did not emerge as a recent fad, but as the long-overdue return of questions that had long been set aside.

Biological non-locality.

One of the most unsettling ideas in contemporary biology is also one of the simplest to articulate: in living organisms, no one is really in charge. There is no control room, no CEO, no command center where the movements of cells, tissues, and organs are decided. Yet the result is right before our eyes: fine-tuned, adaptive, and often elegant coordination. This paradox—order without orders—is at the heart of biological non-locality.

A fascinating example comes from the world of social insects. In 1991, Swiss biologist Laurent Keller observed colonies of *Leptothorax* ants in the Jura forests. When the nest was disturbed, no single ant "decided" where to relocate. Yet, within a few hours, the colony converged on a new site—often the best one available. The choice emerged from a myriad of local micro-interactions, but the final behavior was global. Keller recounted this years later with a quip that has become famous in seminars: "*The colony is smarter than any single ant, but it is not a brain.*"

Something similar happens in multicellular organisms, though it is less visible. During wound healing, for example, cells do not wait for instructions from above. Studies conducted at Cambridge in the early 2000s showed that the healing front organizes itself as a collective migration: each cell "senses" the others, responding to mechanical and chemical gradients that are not tied to a specific point but to the overall state of the tissue. Coordination is not imposed; it is shared.

This idea challenges a deeply ingrained way of thinking: the notion that every function must have a corresponding center. In the human body, we are tempted to attribute this role to the brain. But as early as the 1980s, the French physiologist Alain Berthoz pointed out that the nervous system does not govern like a general, but rather like a diplomat: it negotiates, integrates, and harmonizes signals that arise everywhere. The heart, the gut, and the immune system generate local rhythms and decisions that influence the whole. The result is a sort of biological democracy—unstable but effective.

Biological non-locality should not be confused with a lack of structure. On the contrary, it is a diffuse structure, where information circulates without a single central point. As the philosopher of science Georges Canguilhem wrote, *"life is that which is capable of a* norm," not that which obeys an external norm. Centerless coordination is the concrete form of this autonomy: the organism does not execute a plan; it continuously generates one.

For the contemporary reader, accustomed to hierarchical organizations and decision-making algorithms, this idea has a provocative edge. But perhaps this is precisely the most relevant lesson from field biology: coherence does not arise from total control, but rather from the quality of relationships. And life, from this perspective, remains the most advanced system we know.

Vital Tensions: The Silent Language of Tissues.

There is one aspect of life that has long remained under the radar of molecular biology: the fact that bodies sense forces. Not in a metaphorical sense, but in a physical one. Tensions, compressions, stretches, and the elasticity of tissues constitute a true non-local language, capable of coordinating the organism without the need for specific chemical messengers or decision-making centers. It is a form of communication that travels through the very form of the body.

In the 1990s, in the laboratories of Harvard Medical School, physician and bioengineer Donald Ingber began speaking of *"biological tensegrity*": a structure in which rigid and flexible elements support one another through a continuous network of tensions. The idea, inspired by the sculptures of Buckminster Fuller, initially found more traction among architects than among biologists. Yet the experiments spoke for themselves: by modifying the mechanical tension on a single cell, Ingber observed coordinated changes in gene expression and in the behavior of the entire tissue. Not a local signal, but a global response, distributed throughout the structure.

A fascinating example comes from embryology. As early as the 1980s, in the laboratories at the University of California, Berkeley, Gerald Oster and his colleagues demonstrated that during embryonic development, many forms emerge not because they are molecularly "programmed," but because they are subject to mechanical stress fields. Cells align, migrate, or differentiate along lines of force that traverse the entire embryo. Just as in a Gothic cathedral, where each stone "knows" where to be—not because it receives an order, but because it responds to the tensions of the whole.

An orthopedic surgeon from Lyon recounted that during certain bone reconstructions, the best outcome depended not so much on the millimeter-precise placement of the implant as on the restoration of the correct load-bearing lines. *"If the forces start flowing again, the bone does the rest.*" It is an empirical observation that field biology now interprets more broadly: form informs, and it does so everywhere simultaneously.

This insight is not entirely new. In 1917, the Scottish zoologist D'Arcy Wentworth Thompson wrote in *On Growth and Form* that *"the form of an organism is a diagram of the forces that have acted upon it.*" At the time, it seemed like an elegant but marginal observation. Today, it appears almost prophetic. Mechanical forces are not mere side effects of biochemistry: they are vectors of information that traverse the body— —as a continuous field, enabling coordination without a center.

In this sense, mechanical signals perfectly embody biological non-locality. They do not originate from a single point, nor do they reach an isolated target. They alter the global state of the system, and every part responds as part of that whole. It is a biology that listens to the body in its entirety, even before deciphering its molecules.

Green Thinking: How Plants Coordinate Without a Brain.

For centuries, plants were considered the motionless backdrop of life: passive beings, devoid of behavior, certainly devoid of a mind. A view that says far more about our prejudices than about plant biology. And one that began to crack seriously at the start of the 21st century, when in Florence—not in a high-tech Silicon Valley lab, but in a city steeped in art and history—some experiments showed just how sophisticated plant coordination is, even in the total absence of a nervous system.

In 2005, at the University of Florence, Stefano Mancuso and his team began studying the root tip as a diffuse sensory and decision-making center. Charles Darwin, in a little-cited 1880 text (**The Power of Movement in Plants**), had already intuited something similar, comparing the root tip to the brain of a lower animal. For over a century, however, that observation had remained on the sidelines. Mancuso took it seriously, but without resorting to metaphors: he measured electrical signals, chemical gradients, and growth variations in response to complex environmental stimuli.

The result is surprising: a plant coordinates its behavior as a non-local system, where no single organ makes decisions for the whole and where information circulates throughout the entire organism. If a root encounters an obstacle, the entire root architecture reorganizes itself. If a leaf is attacked, other parts of the plant activate defensive responses even before they are touched. This is not a mechanical reflex: it is a global, coherent response that involves the organism as a whole.

An anecdote shared by a researcher at LINV—the *"International Laboratory of Plant Neurobiology,"* founded in Florence in 2007—illustrates this point well. During an experiment involving competition among plants for limited resources, one plant altered its root growth pattern by anticipating the moves of a neighboring plant. It wasn't reacting; it seemed to be predicting. Not a conscious prediction, of course, but an ability to integrate information distributed throughout the organism's biological field.

Here, biological non-locality becomes evident: the plant does not respond point by point, but state by state. Every environmental variation is translated into a change in the overall information field, which guides growth, flowering, and defense. It is a slow coordination compared to our neural timescales, but no less refined. Indeed, perhaps more robust, because it does not depend on a vulnerable center.

It is no coincidence that Mancuso likes to point out that plants account for over 99% of the planet's biomass and have survived, unscathed, mass extinctions that wiped out "more evolved" organisms. Their strength lies not in speed, but in the distribution of intelligence. This biological lesson resonates even beyond the boundaries of science: in 2018, during a conference in Paris, an urban planner cited plant models specifically as a way to rethink cities that are less centralized and more resilient.

Within the framework of field biology, plants thus become a prime example: organisms that demonstrate how life can be deeply coordinated without a brain, without a center, and without a command structure. Not out of necessity, but by evolutionary choice. And perhaps this is precisely what makes them so relevant today.

Green Blindness: Why the West Has Been Slow to Understand Plants.

For much of Western history, plants have been victims of a sort of ontological prejudice. They lived, of course, but in a gray area of the hierarchy of being: below animals, far removed from humans, too silent to be interesting. Aristotle had already placed them on the lowest rung of the soul—the "vegetative" soul—devoid of sensation and desire. A classification that, as often happens, became a cultural destiny.

While modern science developed by pursuing movement, speed, and observable behavior—everything that could be measured, dissected, or isolated—the plant world remained in the background. Galileo observed the heavens, Vesalius dissected animal bodies, and Harvey traced the flow of blood. Plants, at best, were cataloged. Linnaeus, in the 18th century, constructed a flawless system of classification, but his botany was primarily nomenclature, not an understanding of processes.

This lag was not merely scientific, but deeply cultural. The West has long conceived of knowledge as something that arises from a center—the brain, the subject, the mind—and radiates outward. Within this framework, an organism without a nervous system could only appear simple, reactive, almost mechanical. This view was consistent with Cartesian reductionism but ill-suited to grasping forms of distributed and non-local organization.

Elsewhere, the perspective was different. In Japan, as early as the Edo period (17th–19th centuries), the relationship with the plant world was not based on the idea of domination or classification, but on that of resonance. The art of ikebana, codified centuries earlier, does not consist in "arranging flowers," but in making visible the lines of force between stem, space, and void. In the Zen gardens of Kyoto—take, for example, Ryōan-ji, completed at the end of the 15th century—d nature is not decoration, but a system of relationships: each element makes sense only within the context that encompasses it.

A British anthropologist recounted, after a long stay among the Amazonian peoples, that he had been struck by the way certain communities spoke of medicinal plants: not as resources, but as beings with whom they maintain a relationship. This was neither romanticism nor naive animism, but a refined practical knowledge, based on centuries of observing complex plant behaviors: response times, invisible interactions, slow but consistent adaptations.

The West, on the other hand, took much longer to accept that complexity does not equate to speed or to the presence of a control center. It was only in the second half of the twentieth century, with the crisis of linear models and the emergence of the science of complex systems, that it became possible to seriously reconsider organisms that function "as a field," where the whole precedes and conditions the parts.

It is no coincidence that this reevaluation occurred in parallel with a broader cultural rethinking: ecology, critical anthropology, and even certain currents in contemporary aesthetics began to challenge the idea of an isolated subject observing an inert object. From this perspective, plants have become a conceptual testing ground. Accepting their complexity means accepting that life can be organized without a center, that intelligence can be distributed, and that causality can be non-local.

Perhaps the West's true lag has not been scientific, but imaginative. We have struggled to recognize what did not resemble us. And plants, precisely because they are so radically "other," demanded a paradigm shift that we are only now beginning to address.

Do Plants "Know"? A New Ecology of Relationships.

Recognizing that plants possess a form of distributed intelligence is not merely a scientific act. It is a cultural gesture that shifts the axis of our relationship with the environment. As long as we conceived of the plant world as a collection of inert objects—resources to be extracted, surfaces to be occupied, biomass to be optimized—ecology remained, at best, a more prudent management of exploitation. But if we accept that plants are organisms capable of integrating information, coordinating, and responding coherently as non-local systems, then the question changes: no longer "how much can we take?", but "with whom are we interacting?".

This shift is already evident in the field. In 1997, during a reforestation project near Vancouver, ecologists noticed that plantations established with respect for local plant diversity showed significantly greater resilience than monocultures. It wasn't just a matter of soil or climate. It was the behavior of the system as a whole that was changing: a collective response to environmental stresses. At the time, people spoke of "emergent effects." Today we would say: field dynamics.

The idea that an ecosystem is not the sum of its species, but a stabilized information structure, has practical implications. In agriculture, for example, it is changing the way crop management is conceived. On some biodynamic farms in the Loire Valley, the decision was made to drastically reduce chemical interventions—not out of ideology, but based on observation: less "controlled" fields demonstrated a surprising capacity for self-regulation. Not because plants are benevolent, but because a coherent system reacts better than a fragmented one.

On an ecological level, recognizing the distributed intelligence of plants also means rethinking the concept of environmental damage. Cutting down a forest is not simply a matter of removing trees; it means disrupting a network of delicate relationships that has stabilized over time. In 2019, during the wildfires that devastated large areas of Australia, some local botanical e s emphasized that the loss was not only quantitative but also qualitative: certain ecosystems, once fragmented, do not “recover,” because their field of coherence is broken.

This awareness is also making its way into the law. In 2008, Ecuador enshrined the recognition of the rights of Nature in its Constitution. This act is often dismissed as symbolic, but it stems from a vision—shared by many indigenous Andean peoples—in which the living world is a network of interdependent subjects, not a collection of objects. This is not about attributing human intentions to plants, but about recognizing that acting on an organism or an ecosystem means intervening in a network of relationships.

Even ecological language is changing. Words like “management,” “control,” and “sustainable exploitation” are beginning to reveal their limitations. If plants are complex information systems, then the environment is not a passive backdrop, but an interlocutor. Aldo Leopold had already intuited this when, in 1949, he spoke of the “land ethic”: an invitation to consider the soil, water, and plants as members of a biotic community. Today, that intuition has a more solid biological foundation.

From this perspective, ecology is no longer merely a science of conservation, but a science of coexistence. Recognizing the distributed intelligence of plants compels us to think in terms of systems, long time scales, and diffuse causality.

The Myth of Command: What Human Organizations Fail to Understand About Life.

Whenever a human community attempts to function without a clearly defined chain of command, an almost instinctive mistrust sets in. “Who decides?”, “Who is responsible?”, “Who holds everything together?” Legitimate questions, certainly, but also revealing ones. Because biology has long shown us that highly complex systems can coordinate without a central command, while our social imagination continues to struggle to accept this.

The comparison with non-local organisms is not meant to draw naive analogies—a cooperative is not an organism, a cell is not an assembly—but to highlight a deep-seated cultural limitation: our difficulty in conceiving of coordination as a property of the system rather than as a function of a central authority.

In the human world, attempts at non-hierarchical organization have often been viewed as fragile experiments, destined for chaos or inefficiency. Yet some cases have stood the test of time. The Mondragon cooperative group, founded in the Basque Country in 1956, now involves tens of thousands of workers in a system where ownership and many decisions are distributed. It is not a conflict-free utopia, but a concrete example of how coherence can emerge from shared rules and continuous feedback, rather than from orders handed down from above.

Open-source projects present yet another case. In 2001, during a meeting in San Diego, a developer explained how the software he was working on improved precisely because no one was in complete control. Linux, often mistakenly cited as an anarchic model, actually functions thanks to distributed causality: local contributions, peer reviews, and reputations that emerge over time. There is no "directing molecule," but rather a structure that maintains coherence while constantly changing.

What is striking when observing these models is how culturally demanding they are. The assemblies of the *Occupy Wall Street* movement in New York in 2011 were an emblematic example of this. The absence of visible leaders was not a sign of organizational incapacity, but a deliberate choice. However, many observers—journalists, politicians, academics—interpreted that lack of a central figure as a sign of r immaturity. As if, without a leader, no form could exist.

The biology of fields suggests a different interpretation. In non-local living systems, coordination does not occur because someone "knows more," but because the global state of the system constrains and guides the parts. Applied to the human realm, this implies that norms, contexts, shared values, and information flows matter more than figures of authority. This is a lesson that economist Elinor Ostrom had already grasped while studying, between the 1970s and 1990s, communities capable of managing common-pool resources without central authorities—from irrigation systems in Nepal to Japan's forests.

The critical point is not to determine whether these models are always preferable, but to recognize that our conception of coordination is historically narrow. We have internalized the idea that order must be imposed, not emerge; that coherence requires a center, not a network of relationships. This is why non-hierarchical models seem unstable to us, while biological ones—which function according to similar logics—seem "natural."

Perhaps the true challenge posed by biological non-locality does not concern efficiency, but imagination. Accepting that the whole can come before the parts—even in human organizations—means letting go of a certain idea of control. And this, more than a technical difficulty, is a cultural threshold that we have not yet fully crossed.

When everything influences everything: the end of the sovereign molecule.

Every scientific era has its fetish. For twentieth-century biology, it was long the idea of the decisive molecule: the key gene, the master hormone, the receptor that "explains everything." A reassuring, almost narrative figure. If there's a director, the movie becomes understandable. But life, on closer inspection, doesn't like linear scripts.

The idea that there is a central cause within organisms—a sort of molecular puppeteer—stems more from a cultural need than from empirical evidence. It reflects a mindset that prefers vertical explanations: something commands, something else obeys. Yet the more biology has refined its tools, the more this image has cracked.

In the 1980s, while studying gene regulation in yeast, researchers at the EMBL laboratory in Heidelberg noticed an inconvenient fact: the same genes could produce opposite effects depending on the cell's overall state. There was no "leader gene" that determined the outcome. It was the global context that gave meaning to local action. Causality, rather than descending from above, was distributed laterally.

A striking example comes from contemporary embryology. In 2006, at *the Karolinska Institutet* in Stockholm, a research team observed that identical embryonic cells, placed in slightly different environments, took divergent developmental paths despite having the same genetic makeup. It was not the DNA sequence that had changed, but the network of relationships in which those cells were embedded. As the Swiss bi st Denis

Duboule wrote in a lecture in Geneva: *"The genome is not a musical score that plays on its own; it is a system that responds to the state of the orchestra.*"

This image captures an essential point: in living organisms, causes are not isolated but interdependent. An enzyme functions because the pH, temperature, cellular structure, and overall metabolic state make it possible. Change the environment, and the function changes. Searching for the "master molecule" is tantamount to confusing a single node in the network with the entire network.

Neuroscientist Karl Friston, speaking in London in 2012 about the brain as a predictive system, observed that "causes in living systems are not discrete events, but distributed constraints." A statement that holds true far beyond neuroscience. Life does not proceed by orders, but by conditions.

Curiously, this idea is not entirely new. As early as 1934, the American physiologist Walter Cannon, while studying homeostasis at Harvard, insisted that the body's equilibrium did not depend on a single control center, but on a plurality of interacting mechanisms. The term "feedback"—now overused—was coined precisely to describe this circular causality, where the effect modifies the cause.

Later, in the 1970s, the Italian biochemist Enrico Bucci often told his students that *"the cell does not read instructions; it interprets signals.*" A revealing verb. To interpret implies a context, not a command. Clinical medicine also offers eloquent examples. Some drugs exhibit different—and sometimes opposite—effects depending on the patient's overall condition. It is not merely a matter of dosage, but of systemic organization. The active ingredient does not "decide" anything on its own. It acts within a biological field that modulates its meaning.

If the "director" molecule does not exist, why do we keep looking for it? Because it simplifies things. It reduces uncertainty and promises control. But the price is high: we lose sight of what truly keeps an organism alive—namely, the dynamic coherence of the whole. The philosopher of science Isabelle Stengers, in a public dialogue in Brussels in 2015, observed that *"modern science has often confused explaining with dominating."* Distributed causality, on the contrary, calls for a different attitude: understanding without reducing, intervening without claiming to command.

Accepting that there is no single sovereign molecule does not mean giving up precision, but rather changing one's perspective. It means shifting from the search for a single cause to the interpretation of configurations. And it is precisely this shift that sets the stage for the subsequent chapters of the book, where the concept of the biological field will become inevitable.

We are not born in pieces: because in biology, the whole truly comes first.

There is a persistent misconception that runs through modern biology like an underground current: the idea that to understand an organism, one need only take it apart. Cell by cell, molecule by molecule, gene by gene. An approach that has yielded extraordinary results, to be sure, but one that leaves an awkward question on the table: when exactly do those parts become an organism?

The primacy of the organism is not a philosophical slogan, but an empirical observation that is increasingly difficult to ignore. In living beings, the parts do not pre-exist the whole like ready-to-use building blocks. On the contrary, it is the organism—understood as a dynamic, historical, coherent unit—that makes its parts possible and functional. Without the whole, the parts "mean" nothing.

In the 1950s, the neurologist and psychiatrist Kurt Goldstein, while working with brain-injured soldiers in hospitals in Frankfurt and later in New York, observed a surprising fact: a localized lesion never produced a purely localized deficit. It was the patient's entire behavior that reorganized itself. Goldstein arrived at a conclusion that was uncomfortable at the time: the organism always reacts as a whole. It does not repair a broken part, but rather redefines its own vital norm.

A fascinating example comes from contemporary cell biology. Stem cells, which have been studied intensively since the 1990s in laboratories such as those at Cambridge and the Weizmann Institute, do not possess an intrinsic destiny. They become neurons, hepatocytes, or h r muscle cells not because they "carry within them" a predestined role, but because the organism—or the tissue—places them within a network of relationships that shapes their identity. The same cell, when removed from its context, rapidly loses that capacity. This was already noted in the 1970s by Mina Bissell at Berkeley: breast cells cultured in vitro ceased to behave as such. The DNA had not changed. The organism, as the condition of meaning, was no longer present. Here, the primacy of the whole becomes evident: the parts are not absolute entities, but situated functions. Their "nature" depends on the role they play within the whole.

The immune system, often described as an army of specialized cells, also offers a valuable lesson. Niels Jerne, a Danish immunologist and 1984 Nobel laureate, proposed the idea of an immune network: antibodies that recognize other antibodies, in a game of mutual balance. There is no control center that decides what is "self" and what is "non-self." It is the organism, through its history, that continually establishes this distinction. An anecdote recounted by one of his collaborators in Basel recalls how Jerne liked to say: *"The immune system does not defend the body; it is the body that, by living, shapes its immune system."* Once again, the whole precedes and grounds the parts.

This reversal also has profound implications for the concept of form. As early as 1917, in his **On Growth and* Form*, the biologist and mathematician D'Arcy Wentworth Thompson suggested that form is not the result of specific instructions, but of global constraints. Today we know that those insights were surprisingly prescient.

An organism is not an assembled design, but a self-sustaining process. Parts emerge, persist, or disappear depending on the stability of the whole. This is why, as Georges Canguilhem observed in his lectures at the Sorbonne in the 1960s, biological normality is not a statistical average, but a capacity to establish norms—a capacity that belongs solely to the organism as a whole.

Accepting the primacy of the organism also means changing the way we intervene in life—in medicine, in ecology, and even in biotechnology. It is not a matter of abandoning analysis, but of recognizing that the unit of meaning in living beings is not the part, but rather the organized relationship among the parts.

It is a silent revolution, without manifestos, but with radical consequences. As the theoretical biologist Robert Rosen wrote shortly before his death in 1998 in Halifax: *"An organism cannot be explained by what it is made of, but by what it makes possible.*" A sentence that precisely sums up the heart of this chapter.

PART THREE. Where Life Takes Center Stage.

From the thought of Michael Faraday:

The principles of the great are always present in the small. A phrase born of physics, but one that today resonates surprisingly with biology, when we attempt to understand how the order of the living emerges without a central director.

From the thoughts of James Clerk Maxwell:

At a certain point, it is no longer matter that dominates the scene, but the relationships that run through it. In his studies of fields, Maxwell intuited that reality is not made up solely of objects, but of dynamic configurations that hold them together.

After demonstrating, in the previous parts, the limits of the reductionist paradigm and introducing life as a phenomenon of global coherence, Part Three of the book takes a further and bolder step: it dares to name what has so far remained in the background. Here the concept of the biological field enters the picture—not as a poetic suggestion or a remnant of ancient vitalistic traditions, but as a rigorous attempt to provide an adequate language for describing the organization of life.

Chapter V directly addresses a delicate question: what exactly is a biological field? Not a convenient metaphor to fill the gaps in our explanation, but a conceptual hypothesis arising from the observation of phenomena that are difficult to explain in purely local terms. Speaking of a field means shifting the focus from the isolated object to the configuration that makes it what it is, from the "where" to the "how" of living form. It is a shift in perspective that brings biology closer to what physics has already learned in the twentieth century: that order can emerge without direct contact, that action can be distributed, and that the whole is not simply the sum of its parts.

In Chapter VI, the discussion delves even deeper, focusing on the notion of embodied information. Here, information is not an abstract code floating above matter, nor a simple flow of measurable signals. It is something that takes shape, that expresses itself in space and time, that becomes texture, rhythm, and orientation. Life thus appears as a stabilized informational event: not a message written once and for all, but a process that continually renews itself through the field that sustains it. In this sense, the biological field becomes the place where information ceases to be a given and becomes a behavior.

Finally, Chapter VII addresses a term as ancient as it is controversial: vital energy. Not to resurrect it out of nostalgia, but to critically redefine it. Here, energy is not a mysterious entity opposed to biochemistry, but an organized mode of exchange, of maintaining coherence, and of resisting dispersion. Reinterpreted in the light of fields, vital energy is not something added to matter, but what holds it together in a recognizable living form.

Within the book's overall structure, this Part Three represents a true conceptual turning point. While the previous sections have shown why reductionism is insufficient and how life manifests as global coherence, here we attempt to answer the most challenging question: through which invisible structures does this coherence stabilize and maintain itself? The fields of life are not the conclusion of the discussion, but its point of greatest theoretical density: the place where biology, physics, and philosophical reflection begin to speak a common language.

From this point forward, it will no longer be possible to speak of health, disease, knowledge, and responsibility without coming to terms with this field dimension—which cannot be seen but organizes, cannot be measured directly but guides, does not command but makes things possible. It is the transition in which biology, at last, ceases to merely describe what life is made of and begins to seriously question what makes it life.

V. What Is a Biological Field?

At this point in our journey, the question is no longer whether living beings can be explained as a simple sum of components, but rather with what conceptual tools it is possible to describe their unity. This Chapter V arises precisely from this need: to give a name and a theoretical foundation to that which, within organisms, coordinates, stabilizes, and directs processes without coinciding with a single molecule, reaction, or gene.

The biological field is introduced here as a physical and informational reality, not as an evocative metaphor nor as a remnant of vitalism. It is an attempt to bring into biology a concept that has already proven decisive in other sciences: the idea that order resides not only in objects, but in the space of relationships that makes them possible. Speaking of a field means shifting our focus from linear causality to the overall configuration, from the part to the context that gives it meaning.

In the history of science, the concept of a field emerges whenever explanations based on isolated entities reveal their limitations. A field is not just another thing, but a different way of describing reality: not what acts, but rather that within which action becomes possible. Applied to biology, this concept suggests that the organism is not merely a collection of material components, but a dynamic configuration that maintains its identity through change.

As Albert Einstein wrote, *"a field is an essential part of the physical description of reality, not merely a* mathematical *artifice.*" This statement originated in the context of physics, but it takes on particular resonance when one observes the ability of organisms to preserve form and function even amid the continuous turnover of matter.

One of the central objectives of this chapter is to draw a clear line. The biological realm is not a mysterious force; it is not an occult principle added to chemistry and physics; nor is it a conceptual shortcut to avoid difficult explanations. On the contrary, it is a proposal that demands greater rigor, because it compels us to conceive of causality in distributed, rather than local, terms.

In this sense, Richard Lewontin's warning resonates, when he observed that *"the causes of biological phenomena never lie at a single level of organization.*" The biological field does not replace known mechanisms, but rather organizes them, constrains them, and delimits their scope of expression.

The theoretical heart of the chapter lies here: the biological field is described as a structure of possibilities. It does not determine every event, but makes some events more probable than others; it does not impose a rigid form, but stabilizes a coherence. This is what allows an organism to develop reliably without being rigidly programmed, to respond to the environment without dissolving into chaos.

This idea finds a surprisingly clear formulation in the words of Kurt Lewin, one of the first to explicitly use the concept of a field in living systems: *"Behavior is a function of the person and his or her field.*" Transposed into biology, this statement suggests that the form and function are not isolated internal properties, but rather emergent properties of a continuous interaction between structure and context.

Within the book's overall structure, this chapter serves as a crucial transitional point. After demonstrating, in the preceding sections, the crisis of reductionist explanations and the importance of coherence as a hallmark of life, this chapter introduces the language necessary to move beyond them. The biological field becomes the bridging concept that allows us to discuss, in subsequent chapters, embodied information and vital energy without slipping into either mechanistic or mystical interpretations.

Biological field and environment: a necessary distinction.

When we speak of the biological field, misunderstanding is always lurking: isn't this perhaps just an elegant way of referring to the environment? The soil, temperature, light, and the chemical and social context surrounding an organism? The confusion is understandable, but it is precisely here that one of the most important distinctions of this new vision of life comes into play. The biological field is not the same as the environment, just as music is not the same as the air that carries it.

The environment is what surrounds an organism: a set of external, variable, and often contingent conditions. It can foster or hinder life, nourish it or threaten it. The biological field, on the other hand, is intrinsic to the organism, even if not rigidly confined within its physical boundaries. It is an organizational reality that precedes and coordinates interactions with the outside world. The environment proposes; the field disposes.

A simple yet illuminating example comes from early 20th-century experimental botany. In Vienna during the 1920s, the physiologist Gottlieb Haberlandt observed how isolated plant cells could regenerate entire tissues when placed under the right conditions. The environment provided nutrients and stimuli, of course. But the form that emerged—leaf, stem, root—was not a pre d in the soil or nutrient solution. It was as if the plant "knew" what to become. Today we would say: it followed an internal structure of possibilities, not deducible from the environment itself.

The biological field is precisely this: a matrix of organization, an informational structure that guides vital processes without being reducible to either genetics or the external context. The environment acts as a set of boundary conditions; the field as a principle of integration. Confusing the two is like mistaking the stage for the script.

An analogy often used by complex systems scientists themselves—and not by passing mystics—is that of an orchestra. The concert hall, with its acoustics, influences the performance: it is the environment. But the symphony does not arise from the hall. It is the orchestra's internal organization, the shared tempo, and the coherence among the instruments that produce the music. The biological field is more akin to this dynamic coherence than to the physical space that hosts it.

This distinction becomes crucial when we observe organisms capable of maintaining a stable form and behavior in extremely diverse environments. Consider tardigrades, which have been studied since the 18th century and became famous in the 21st century for their resilience: they survive the vacuum of space, radiation, and extreme cold. The environment may vary radically, but the organism retains—in a latent form—its own organization. It is difficult to explain this continuity by appealing solely to external conditions. It is much more plausible to conceive of a biological field that can contract, suspend itself, and reactivate without dissolving.

Stating that the field is not the environment also helps to curb pseudoscientific excesses. Not everything that "influences" an organism constitutes its field. City traffic, social stress, and light pollution have real, documented biological effects. But they remain environmental factors. The biological field is not the sum of external stimuli, nor is it a vague cloud that absorbs everything that happens around a living being. It is a specific structure, with its own internal coherence, that interacts with the environment without dissolving into it.

In this sense, the biological field can be understood as an active memory of form, not localized at a precise point but distributed throughout the organism. The environment can disrupt this memory, facilitate it, or hinder it, but it does not generate it. It is the difference between weather conditions and climate: the former change constantly, while the latter emerges from long-term consistency.

Conceptually separating the field and the environment also has a significant cultural implication. It means recognizing that life is not merely passive adaptation to external events, but also the assertion of an organized inner life, capable of responding without being completely determined. In an era that tends to explain the human—and living beings in general—as a product of circumstances, this distinction reopens a space for biological autonomy, even before ethical autonomy.

The biological field, therefore, is not what surrounds us, but what holds us together as we navigate what surrounds us. The environment changes; the field persists. The environment stimulates; the field organizes. To confuse them is to lose sight of the very heart of the phenomenon of life.

Inhabiting the world, not merely enduring it: from the concept of "Umwelt" to the biological field.

To fully understand what a biological field is—and what it is not—it is helpful to take a step sideways, rather than backward, and listen to a philosophical voice that twentieth-century biology has long engaged with without ever fully absorbing it: that of Jakob von Uexküll. A Baltic-German biologist and thinker active from the late nineteenth century through the

1930s, Uexküll introduced an idea as simple as it was unsettling: every organism lives within its own *Umwelt, a "surrounding* world" *that* is selected, meaningful, and tailored to its senses and capacities for action.

The famous example of the tick, discussed in his **Environment and Behavior of* Animals* (1934), has become almost a rite of passage for generations of scholars. The bat, suspended from a branch, responds to only three signals: the smell of butyric acid emitted by mammals, their body heat, and the texture of their skin. Everything else—the forest, the colors, the sounds—simply does not exist. Its world is extremely limited, yet perfectly suited to its life.

Here lies a first, important affinity with the concept of the biological field: life does not react to an objective and neutral environment, but to a world structured by meaning. However, stopping at this similarity risks creating a new misunderstanding. Uexküll's *Umwelt* is above all a perceptual and functional construct, a bubble of meanings that emerges from the encounter between organism and world. The biological field, on the other hand, operates at a deeper level: it defines not only what the organism perceives, but what the organism is capable of becoming.

In other words, the Umwelt describes the world as it appears to a living being; the biological field describes the internal coherence that makes that appearance possible. One is phenomenological; the other is organizational. The Umwelt changes if the senses or functions change; the biological field changes only when the organism's own informational stability is altered.

A little-known anecdote, recounted by one of Uexküll's students, concerns a discussion that took place in Hamburg in 1928 during an informal seminar. When asked whether the Umwelt was a property of the world or of the organism, Jakob von Uexküll replied with a statement that sounded almost provocative: *"The world is a musical score, but each organism plays only its own melody*." A powerful metaphor, yet one that leaves a crucial question unanswered: who holds the melody together over time?

This is where the concept of the biological field comes into play without overlapping. The field is not the perceived melody, but the structure that makes performance possible, maintaining coherence even when conditions change. Two organisms may share a similar environment and even a partially overlapping Umwelt, yet have profoundly different biological fields—as occurs, for example, in different species raised in the same artificial context, or in individuals of the same species with divergent developmental histories.

This distinction has also resonated outside of biology. The philosopher Maurice Merleau-Ponty, in his lectures at the Collège de France between 1956 and 1959, drew on Uexküll to criticize the idea of an organism as a reactive machine. But he went further, suggesting that the living form possesses an "internal logic" that cannot be reduced to either stimulus or perception. It is a subtle yet decisive shift: from the Umwelt as the lived world to the field as the organizing principle of life.

The implicit dialogue between Uexküll and field biology is therefore not an endorsement, but a line of critical continuity. Uexküll taught us that there is no single world for all living beings. The theory of biological fields adds that there is also no purely external explanation for the form and stability of life. Every organism not only inhabits a world but carries within itself the rules of its own habitability.

In this sense, the biological field does not replace the Umwelt, but presupposes and transcends it. If the Umwelt is the world as it appears to a living being, the field is what allows that living being to have a world, to maintain its coherence, and to reconfigure it without dissolving. A subtle but decisive difference that marks the transition from a biology of perception to a biology of meaning.

The form that sustains itself: development, stability, and fields of life.

One of the oldest questions in biology is also one of the most disarming in its simplicity: who decides the form of an organism? Who "tells" one cell to become part of a liver and another to transform into a neuron? For a long time, the implicit answer was reassuring: somewhere there must be a control center, a program, a molecular control room. But the more biology has examined development closely, the more this center has begun to fade away.

The idea of the biological field arises precisely from this: from the recognition that the stability of living forms does not require central control, but emerges from a distributed, coherent regulation that is continuously negotiated among the parts. There is no conductor, yet the symphony goes on.

A historical episode often recounted in the corridors of embryology departments dates back to the 1920s, at the Kaiser Wilhelm Institute in Berlin. The embryologist Hans Spemann, a future Nobel laureate (1935), demonstrated that a salamander embryo, even when partially damaged or reorganized, was capable of developing in a surprisingly orderly manner. By cutting, moving, or even rotating portions of embryonic tissue, the final form still tended to reassemble itself. Not perfectly, of course, but recognizably so.

The crucial point was not "plasticity" per se—a term that has since become overused—but rather the fact that the embryo seemed to know where it was going, even when the initial conditions were altered. Spemann spoke of "morphogenetic fields," an expression that at the time sounded more descriptive than explanatory, but which we can now reinterpret as a fundamental insight: development is not a linear chain of instructions, but rather a process regulated by global constraints.

This regulation does not function like an urban plan imposed from above, but rather like the self-regulation of a historic city.

Anyone who has walked through Venice knows this: there is no perfect symmetry, yet the whole holds together. The streets adapt, spaces balance each other out, and functions emerge from use. Similarly, in biology, each cell responds to local signals, but it does so within a configuration of possibilities that limits and guides its choices.

In the 1970s, developmental biologist Lewis Wolpert proposed the famous "gradient" model, illustrated with the image of the French flag: identical cells take on different fates depending on their position. The idea was hugely successful, but it left a crucial question unanswered: who maintains the stability of the gradient? Who ensures that, despite noise, errors, and fluctuations, the overall pattern does not collapse?

This is where the concept of the biological field demonstrates its explanatory power. The field does not issue orders; it establishes relationships. It does not tell individual cells what to do in a deterministic way, but rather defines the space of possible responses, making some developmental trajectories more likely and others practically inaccessible. Form is not imposed; it is attracted.

A lesser-known but illuminating example comes from studies conducted in Heidelberg in the early 2000s on three-dimensional cell aggregates, now called organoids. When stem cells are allowed to interact without detailed external guidance, they often generate surprisingly orderly structures: layers, cavities, and rudimentary symmetries. Not because they "know" what they are supposed to become, but because the field of interactions that arises among them channels development toward stable configurations.

In this sense, the stability of form is not rigidity. It is resilience. It is the ability to absorb disturbances without losing identity. An effective biological field does not eliminate error: it integrates it

This perspective has profound implications. If development is regulated by a field rather than by a center, then living form is the result of dynamic coherence, not of a rigid, pre d plan. The organism does not execute a plan: it maintains its form as it changes. And it is precisely this capacity to maintain itself—rather than to obey—that is the hallmark of life.

Attractive forms: why life chooses some paths over others.

When we observe the development of an organism—from a single cell to a complex form—the temptation is always the same: to imagine a sequence of instructions carried out step by step. Yet anyone who has had the opportunity to work with real living systems knows that development resembles less a recipe and more a descent through a landscape. Some paths are slippery; others naturally lead toward deep valleys. It is here that mathematical biology has introduced a powerful and surprisingly concrete metaphor: that of attractors.

The term originated in the context of dynamical systems, at the intersection of physics and mathematics, but found a natural home in twentieth-century biology. An attractor is a state toward which a system spontaneously tends, even when starting from different initial conditions. In the 1940s, the mathematician and biologist Conrad Hal Waddington, while working at Cambridge, proposed the now-famous image of the epigenetic landscape: a ball rolling along an undulating surface, entering increasingly well-defined valleys. Each valley represents a stable form, a possible outcome of development.

Waddington drew these images by hand, often on chalkboards grayed by chalk, during lively discussions with colleagues and students. They were not merely educational illustrations: they were a way of saying that form is not constructed piece by piece, but emerges because some configurations are inherently more stable than others. In modern terms, we might say that the biological field creates the landscape, while the attractors are its valleys.

This helps clarify a crucial point in the chapter: not all forms are equally likely. In a developing organism, some architectures are "easy," while others are nearly unattainable. Not because they are prohibited by an external law, but because the field of interactions—genetic, biophysical, temporal—makes certain solutions extremely robust. It is the same principle by which, as the physicist Ilya Prigogine observed in Brussels in the 1960s, systems far from equilibrium do not produce indiscriminate chaos, but preferential order.

An anecdote recounted by a researcher at the Institut Pasteur in the early 1990s illustrates this point well. During experiments on artificially perturbed genetic networks, the system always "returned" to a few recurring states, as if it were ignoring many of the theoretical possibilities. *"It was like pushing a marble out of a trough and watching it stubbornly roll back to the same spot.*" Those points were attractors: stable states defined not by a command, but by the overall geometry of the system.

In the language of field biology, attractors are not abstract entities: they are embodied forms, configurations toward which the organism tends because the biological field renders them coherent. This explains a phenomenon familiar to those who study comparative morphology: different species, even those phylogenetically distant, can converge on similar forms. It is not only natural selection that explains these convergences, but the fact that certain morphological solutions occupy deep regions of the attractor landscape.

A cultural parallel helps to clarify the idea. Art historian Ernst Gombrich observed that, across different eras and places, certain figurative patterns tend to reemerge. Not because artists copy one another, but because some visual solutions "work" better than others. Similarly, in biology, the field does not impose a form, but favors those that persist, that resist fluctuations, that absorb error without losing their identity.

From this perspective, the biological field can be seen as the context that defines the landscape of attractors. Not a program, not a mysterious force, but a structure of possibilities that makes certain trajectories of development more natural, more accessible, almost inevitable.

From order to meaning: how the field transforms information into living form.

At this point in our journey, the biological field can no longer be thought of as a mere backdrop, nor as an elegant idea useful for tying together phenomena that are difficult to explain. If the field is real, if it structures the organism's possibilities and stabilizes certain forms over others, then the question becomes inevitable: what kind of information does the field convey?

Not the passive information of an archive—consisting of data waiting to be read—but something closer to what physicist Donald MacKay called, as early as the 1950s in London, "effective information": information that makes a difference, that guides a process. In biology, the field does not "contain" messages the way a library contains books; rather, it modulates the system's behavior over time, making some responses likely and others unlikely.

An anecdote recounted in the hallways of the *Marine Biological* Laboratory in Woods Hole, Massachusetts, helps clarify this conceptual shift. In the 1980s, while observing seemingly identical cell cultures, some researchers noticed that the same perturbation produced systematically different outcomes depending on the system's history. There was no new genetic information at play, nor were there any obvious environmental variations. What was changing was the overall state of organization: what we would today call distributed information, embedded in the field of relationships.

Here, the biological field reveals its most intriguing aspect: it does not instruct, but rather guides. It does not command, but rather sets the stage. It is information that operates through constraints and possibilities, not through explicit orders. It is somewhat like what happens in music: the key of a piece does not tell the musician which note to play, but it makes some sequences sound natural and others jarring. In a 1969 conversation in Berlin, composer György Ligeti spoke of *"fields of sonic* tension" within which music takes shape. The metaphor is not biological, but the conceptual mechanism is surprisingly similar.

Thinking of the field as active information also helps dispel a common misconception: that biological information is synonymous with DNA. The genome provides necessary, but not sufficient, conditions. It is the field that transforms those potentials into concrete trajectories, actualizing the information over time. As theoretical biologist Howard Pattee noted in the 1970s, information becomes biologically relevant only when it is embedded in a system capable of interpreting and using it. The field is precisely this interpretive context.

In this light, the organism appears as a stabilized informational structure—not because it is immobile, but because it is capable of maintaining coherence while changing. Information is not what the organism possesses, but what happens through it. The field, then, is not an additional entity, but rather the mode through which information becomes effective, corporeal, and situated.

This conceptual shift naturally sets the stage for the next chapter. If the field is information in action, the next step must be to ask how information takes shape, how it becomes tissue, gesture, and function. No longer the "what" of information, but the "how" of its embodiment. It is there that the biology of fields naturally encounters a new idea of life: not as a machine that executes instructions, but as a process that gives meaning to its own possibilities.

From the magnet to the organism: a brief history of an idea called the field.

To understand what we mean today by a biological field, we need to take a step back—or rather, more than one—and return to the moment when science began to suspect that reality was not made up solely of objects, but also of relationships extending across space and time. The concept of a field arose precisely from this intuition, and not in biology.

In the early nineteenth century, in the makeshift and often cold laboratories of industrial Europe, physics was grappling with an perplexing problem: how to explain action at a distance. When Michael Faraday, the son of a London blacksmith and a brilliant self-taught scientist, observed the lines of iron filings arranging themselves around a magnet, he had the impression that space itself had become active. There was no direct contact, no invisible gears: there was something that structured space. Faraday spoke of lines of force, and to many of his contemporaries, he seemed more like a poet than a scientist.

It was James Clerk Maxwell, a few decades later, who gave that intuition mathematical rigor. With his equations, published in the 1860s, the electromagnetic field ceased to be a conceptual device and became an autonomous physical reality. It is no longer merely a way of describing the interaction between charges: it is something that exists, propagates, and carries energy. As Einstein would later write, *"the field is the fundamental physical reality."*

This shift is crucial. With the field, science began to conceive of nature no longer as a collection of isolated objects, but as a continuous web of influences, in which what matters is not just what is there, but how it is distributed. In the 20th century, this idea became more radical: quantum mechanics introduced fields that describe not only what happens, but the probabilities of what can happen. In Göttingen and Copenhagen, between the 1920s and 1930s, the field became a structure of possibilities even before it was a structure of events.

The leap toward complex systems occurred when this logic was extended beyond fundamental physics. In the second half of the 20th century, scholars from various disciplines began using the concept of a field to describe collective phenomena: from atmospheric dynamics to ecological networks, and even social systems. At the Santa Fe Institute, founded in 1984 in New Mexico, physicists, biologists, and economists share a conviction: certain behaviors emerge not from the parts taken individually, but from the system's overall configuration.

Here, the field definitively ceases to be an exotic entity and becomes a rigorous way of conceptualizing organization. A field is not a thing, but a structuring condition: it defines constraints, distributes influences, and coordinates dynamics without the need for a command center. It is a notion that paves the way for field biology, because it shows how it is possible to explain order and stability without resorting to either an external designer or a sum of local causes.

An anecdote recounted by an engineer involved in the early studies of urban traffic in Stuttgart in the 1970s is illuminating. By slightly modifying the timing of traffic lights—without changing the number of cars or the roads—the entire flow of the city changed. *"We didn't act on the vehicles but on the field of possibilities.*" It is the same principle that, when applied to biology, allows us to conceive of the organism as immersed in a field that guides vital processes even before determining them.

When the concept of a field enters the life sciences, it brings with it this legacy: the idea that causality is not merely linear, that space and time matter, and that the organization has a reality of its own. Speaking of a biological field therefore means placing oneself within a solid scientific tradition—one that stretches from Faraday to complex systems—and that invites us to view life not as an assembly of parts, but as a form of extended, dynamic, and informed order.

Does a field really exist? Scientific realism put to the test by the invisible.

When the concept of a field became firmly established in science, it brought with it not only new equations or more elegant models, but also raised a question that spanned the entire 20th century: Does the field truly exist, or is it merely a useful way of talking about the world? This is a less abstract question than it seems, because it touches on the heart of scientific realism—namely, the idea that scientific theories describe, at least in part, the actual structure of nature.

The debate intensified especially between the 1930s and 1960s, when field theory became the dominant language for describing electromagnetism, gravity, and, later, fundamental interactions. In Vienna, Berlin, and Cambridge, philosophers and scientists questioned what they were actually doing when they spoke of fields. Moritz Schlick, a central figure in the Vienna Circle, tended to view them as powerful conceptual tools, but lacking ontological autonomy: what mattered to him was a theory's ability to organize observable experience.

Albert Einstein took the opposite stance; in a 1952 letter to Max Born, he wrote that the field was not a mathematical fiction, but *"the only physical reality worthy of the name."* For Einstein, reducing the field to a convenient language meant losing what physics had painstakingly achieved: the idea that reality does not necessarily coincide with what is directly observable.

This tension runs through the entire philosophy of science of the twentieth century. Karl Popper, while critical of any metaphysics masquerading as science, defended a form of "conjectural" realism: fields are not observable, but are bold hypotheses that withstand attempts at falsification. Bas van Fraassen, on the other hand, in the 1980s, with his constructive empiricism, proposed a more cautious position: we are not obliged to believe that fields exist; it is sufficient that the theories that use them are empirically adequate.

An anecdote recounted by a student who attended a seminar at Princeton in 1967 captures the atmosphere of the time well. After a heated discussion on the ontology of quantum fields, an elderly physicist reportedly closed the debate by saying, *"I don't know if the field exists, but I know that without it, I no longer know how to work."* It's a joke, but it captures an essential point: the field imposes itself as a conceptual necessity, even before it is a metaphysical entity.

This philosophical background is fundamental to understanding the shift toward field biology. When we speak today of a biological field, the question is not whether we are introducing something "mysterious," but whether we are recognizing a real structure of living organization, analogous—in epistemological terms—to the fields of physics. The lesson of the twentieth century is clear: many scientific entities were first tools, then accepted realities. Atoms, genes, and even quarks have crossed this threshold.

From this perspective, the field is neither a material object nor a mere linguistic convention. It is a theoretical entity with explanatory power, whose reality is measured by its ability to make otherwise fragmented phenomena intelligible. Applied to biology, this approach calls for a form of sober realism: the biological field should not be reified, nor should it be reduced to a metaphor. As is often the case in the history of science, its "existence" coincides with its fruitfulness.

Because "field" never goes out of style: a word that captures complexity.

There are words that age rapidly, crushed by the progress of the theories that gave rise to them. And there are others that, on the contrary, seem to span decades—sometimes centuries—changing context without losing precision. "Field" definitely belongs to this second category. It is a word that science continues to use not out of nostalgia, but out of necessity. And for those involved in field biology, this is not a linguistic detail: it is an epistemological clue.

The term entered modern science with a certain timidity. Michael Faraday, a brilliant self-taught scientist who in 1831 was working in the basement of the Royal Institution in London, spoke of *"lines of* force" rather than fields in the subsequent mathematical sense. He literally drew what could not be seen: iron filings that arranged themselves in elegant patterns around a magnet. It was a visual, almost artisanal language. But even then, something decisive was already glimpsed: the idea that space was not a neutral void, but rather a structured region, a carrier of relationships.

Since then, "field" has become a boundary-pushing term. It does not denote an object, does not coincide with a substance, and cannot be reduced to a point force. Rather, it is a way of describing reality that allows us to speak of continuity, distributed influence, and centerless organization. And it is precisely this controlled flexibility that explains its extraordinary longevity.

An anecdote recounted in the 1970s in the hallways of MIT by a young researcher in systems theory is emblematic. During a heated discussion about the terminology to use in describing certain collective behaviors, someone proposed abandoning the term "field" because it was "too vague." The curt reply came from an applied mathematician: *"It's not vague; it's general.*

And generality, when well-grounded, is a virtue." The word stuck.

The point is that "field" works well because it maintains a rare balance between imaginability and rigor. It evokes a space of possibilities—a farm field, a field of vision, a field of action—but at the same time lends itself to precise formalization when needed. It is no coincidence that the word spans distant disciplines: from physics to biology, from Pierre Bourdieu's sociology (who in the 1990s spoke of social fields with surprising conceptual accuracy) to the cognitive sciences. Each time, what fills the field changes, but the logic underpinning it remains the same.

In the biological context, this linguistic persistence is anything but innocent. To speak of a biological field is to choose a word that does not promise more than it can deliver, yet does not reduce life to a list of components either. "Field" suggests a distributed, historically stabilized reality, capable of guiding processes without controlling them. It is a word that leaves room for time, form, and regulation, without succumbing to the temptation of a rigid mechanism.

The linguist Émile Benveniste observed, in a course at the Collège de France in 1969, that certain scientific words survive because "*they organize experience even before explaining it.*" "Field" is one of these. It describes not only what we know, but also how we are inclined to search. And perhaps this is why, whenever science is confronted with complex phenomena—from embryonic development to collective dynamics—that word returns, discreet yet indispensable.

In this sense, the field is not merely a concept: it is a linguistic gesture that reflects a methodological choice. Preferring the term "field" means accepting that biological reality cannot be fully captured in isolated entities, but requires a language capable of holding together relationships, constraints, and possibilities. A word that endures for so long, perhaps, does so not out of inertia, but because it continues to express something true.

Why a "field," specifically? It is the right word to describe organized life.

When biology found itself, during the twentieth century, grappling with phenomena that defied the sum of their parts—the development of a form, the stability of an organism, the capacity for regeneration—it often sought new metaphors. But the field is not a metaphor that arose by chance: it is a concept that science had already refined elsewhere, and one that seemed tailor-made to describe distributed organizations—without a command center, yet far from chaotic.

The concept of the field originated in physics to solve a concrete problem: explaining how one body can influence another at a distance without resorting to mysterious forces. But in doing so, it introduced a more general idea: action no longer passes from one object to another, but rather through a continuous spatial structure, which makes certain configurations possible and others impossible. It is this logic—rather than the mathematical details—that makes the concept surprisingly well-suited to biology.

A developmental biologist observing an embryo does not see a sequence of instructions executed in order, but rather a process in which cells "know" where to go and what to become without any single cell possessing the complete map. It is a situation that resembles, more than is generally acknowledged, that described by a physical field: local conditions guided by a global organization.

The crucial point is that the field allows us to conceive of the organism as a unit extending across time and space. Unlike concepts such as "program" or "machine"—which are too closely tied to a sequential and centralized logic—the field suggests an emergent order capable of sustaining itself even when certain parts are disturbed. It is no coincidence that in the early studies on complex systems conducted at Santa Fe and Los Alamos in the 1980s, the language of fields reemerged to describe robust collective dynamics, in which form persists despite the continuous turnover of components.

There is also a more subtle cultural reason. The concept of a field is intrinsically relational: it does not describe "things," but rather arrangements, gradients, and constraints. And this is precisely what seems to be missing when one attempts to describe living systems solely in terms of molecules. An organism is not made up only of what it contains, but of how what it contains is held together. The field then becomes the name for this cohesion, for this dynamic coherence.

In 1973, during a conference in Heidelberg on morphogenesis, a physicist working in biology observed that the field had a decisive advantage: *"It does not say who is in charge, but what is possible.*" This statement clarifies why this concept, more than others, lends itself to describing living systems. Life does not follow orders; it explores stabilized possibilities.

Speaking of a biological field, at this preliminary stage, does not therefore mean introducing an exotic entity, but rather recognizing that the organization of living beings resembles a structure of active possibilities more than a linear mechanism. The field is the concept that best holds together form, stability, and adaptability without reducing one to the other. And that is why, when biology seeks a word capable of encompassing the complexity of life, it ends up—once again—returning to this concept.

What a biological field is not.

Every powerful concept runs a risk: that of being overused and misused. It happened with the term "energy," it happened with "quantum," and inevitably it happens with "field" as well. In biology, where the complexity of living systems defies simple explanations, the biological field can become a catch-all term: evocative, reassuring, but dangerously ambiguous. Clarifying what a biological field is not is not a defensive exercise, but rather an act of scientific precision.

First and foremost, a biological field is not a mysterious force acting on the organism from the outside. It is not an entity that "envelops" the body like a colored aura, nor is it an invisible presence endowed with its own intentions. The idea of a vital principle hovering over matter belongs to a long cultural tradition—one that is historically respectable but cannot be confused with an operational scientific concept. As the French physiologist Claude Bernard noted as early as 1865, *"invoking a vital force does not explain phenomena; it merely defers them.*"

The biological field is not even a measurable form of energy like a fluid that can be stored or transferred at will. Here, the misunderstanding often stems from the everyday language of science. In a famous lecture at Caltech in 1961, physicist Richard Feynman cautioned his students: *"Energy is not a thing; it is a number that remains constant.*" Applying this lesson to biology means avoiding the notion of the field as a "something" that flows, is , or is consumed. The field is not the fuel of life: it is a mode of organization.

Another common source of confusion concerns its relationship to quantum physics. The biological field is not a quantum field in disguise, nor does it require invoking entanglement or wave collapses to explain every property of living things. In the 1970s, physicist Wolfgang Pauli—in a private letter now preserved at CERN—lamented the overuse of the term "quantum" outside its valid domain. He was right: invoking the quantum becomes pseudoscientific when it substitutes allusion for argumentation.

This does not mean denying the physical foundations of life, but rather recognizing that the biological field operates at a different descriptive level, where the stability of forms, collective regulation, and functional coherence matter. Confusing it with a direct quantum phenomenon is like using a microscope to find your way around a city: a fascinating tool, but unsuitable for the purpose.

It is also important to distinguish the biological field from a mere linguistic metaphor. It is not an elegant way of saying "context" or "internal environment." If that were the case, it would add nothing to our understanding of phenomena. To have scientific merit, the field must entail observable consequences: stability of forms, resilience, and limits on possible variations. When it remains merely an evocative word, it ceases to be a concept and becomes rhetoric.

Finally, the biological field is not an epistemological shortcut to avoid experimental work. It does not serve to fill gaps in knowledge with vague assertions. On the contrary, it is a demanding concept: it requires models, predictions, and comparisons with data. As the British biologist Peter Medawar observed in 1967, *"bad explanations are those that explain everything and therefore nothing*."

Paradoxically, clarifying what a biological field is not serves to reinforce its scope. Freed from mystical encrustations and linguistic abuses, the field can return to being what it promises: a rigorous conceptual tool for thinking about the organization of life without betraying its complexity.

From Equilibrium to Sign: Slow Histories of Difficult Ideas.

The history of biology does not proceed solely through sudden discoveries, but also through uncertain words—concepts that linger for a long time on the threshold between saying and understanding. Ideas that seem vague, sometimes suspect, before becoming indispensable. The field of biology belongs to this lineage. To understand its possible future, it is worth looking at two illustrious examples that endured decades of ambiguity before finding a rigorous place: homeostasis and epigenetics.

When Walter Bradford Cannon coined the term "homeostasis" in 1926 at Harvard, his intent was clear but the ground was slippery. Describing organisms' ability to maintain relatively stable internal conditions seemed, to many physiologists, an elegant way to mask ignorance about the actual mechanisms. In a review published in *Science* in 1933, a colleague observed with irony that homeostasis *"risked becoming a new name for an old mystery."* He was not entirely wrong: for years, the concept oscillated between explanation and description, between an organizing principle and a simple empirical observation.

Yet, over time, homeostasis proved fruitful precisely because it could not be reduced to a single mechanism. It was neither a pump nor a specific circuit, but a logic of distributed regulation. Its strength lay in unifying diverse phenomena—from blood glucose to body temperature—under a common principle. Only later, with the development of cybernetics and feedback systems, did that concept find the appropriate tools to be formalized without losing its generality.

Epigenetics followed a similar, yet even more turbulent, path. When Conrad Hal Waddington spoke of the *"epigenetic* landscape" at Cambridge in the 1940s, drawing hills and valleys to represent embryonic development, many geneticists accused him of using metaphors instead of genes. His language seemed closer to art than to hard science. Yet that metaphor contained a decisive insight: that between genotype and phenotype there existed a regulatory space that could not be reduced to the DNA sequence.

For decades, epigenetics remained a no-man's-land. Too concrete for philosophy, too theoretical for the nascent field of molecular genetics. It was only at the end of the 20th century, with the discovery of DNA methylation mechanisms and histone modifications, that the term was "rehabilitated." But at a price: narrowing its meaning. Many biologists today use "epigenetic" in a technical, almost chemical sense, forgetting the original scope of Waddington's idea. A paradoxical fate: to become acceptable, the concept had to lose part of its ambition.

The field of biology occupies a similar, though not identical, position. Like homeostasis, it concerns the organism's overall stability; like epigenetics, it addresses the issue of form and regulation without reducing them to local instructions. But, unlike both of these, the field did not arise to fill a specific mechanistic void. Rather, it emerged as an attempt to name an organizational dimension that comes to light when linear explanations reveal their limitations.

History suggests caution and patience. Concepts that straddle disciplinary boundaries do not become clear quickly. They go through phases of misuse, rejection, and forced reduction. An anecdote recounted by science historian Evelyn Fox Keller recalls that, during a seminar at MIT in the early 1980s, a researcher defined epigenetics as *"a fancy word for what we don't yet know how to measure*." Today, that very word lies at the heart of developmental biology and medicine.

The decisive difference lies in how these concepts are treated. When used as rhetorical stopgaps, they become fragile. When, on the other hand, they are treated as open-ended problems—capable of guiding research without claiming to resolve it—they can mature. The field of biology, like homeostasis and epigenetics before it, requires time, adequate tools, and rigorous linguistic discipline. It does not promise quick solutions, but rather a new grammar for describing living organisms.

The field as a structure of possibility.

There is a paradox that has always accompanied the study of living organisms: their reliable regularity coexists with surprising elasticity. An embryo follows recognizable stages, yet it can correct errors; a tissue maintains its identity, but can remodel itself; an organism resists disturbances without becoming rigid. The concept of the biological field as a structure of possibility arises precisely from this: from the attempt to describe how life maintains its form without being confined to a rigid schema.

The idea is not that the organism "contains" a form already written out in every detail, but that it operates within a structured space of possibilities, where some configurations are stable, others improbable, and still others simply impossible. The field, in this sense, does not impose specific outcomes: it delimits, orients, and renders plausible. It is a grammar rather than a script.

An episode often cited in developmental biology seminars—and one that garnered some attention as early as the 1960s—concerns the salamanders of the genus Ambystoma studied in the laboratories at Indiana University. When a limb is amputated, regeneration does not produce a random mass of cells, but rather reconstructs a limb that is coherent in terms of size and proportions. Not because each cell "knows" what to do in isolation, but because the entire system operates within a framework that stabilizes certain formal solutions and excludes others. A hand does not grow in place of a limb, nor does an infinite limb. Form emerges as a probable outcome, not as a central command.

Here, the notion of stability must be understood with caution. It is not immobility. Rather, it is dynamic robustness. An organism remains itself not because it does not change, but because it changes within limits that preserve its organization. In 1975, during a conference at Woods Hole, cell biologist Paul Weiss remarked with a quip that has remained famous among experts: *"Life does not defend a state; it defends a relationship.*" It is a phrase that captures well the meaning of the field as a structure of possibility: what remains stable is not matter, but the coherence of relationships.

Biological regulation, too, when viewed from this perspective, loses its character as a "cascading" control mechanism. There is no hidden conductor in the genes or in the embryonic brain. Regulation occurs because the field makes certain deviations self-correcting. When a part deviates too far, the organism does not have to "decide" what to do: it naturally returns to a compatible configuration. This behavior is more reminiscent of architecture than of precision engineering. As in the great Gothic cathedrals—Chartres is often cited by art historians for this reason—stability arises from the distribution of forces, not from a single miraculous pillar.

Talking about the field as a structure of possibilities also helps avoid a common misunderstanding: that of confusing form with an aesthetic or geometric result. In biology, form is always functional, but not in an immediate, utilitarian sense. It is functional because it is sustainable over time, because it can withstand energy fluctuations, molecular noise, and environmental stress. The field selects forms that "stand on their own" biologically, even before they are advantageous in an evolutionary sense.

In this framework, stability is not a brake on innovation, but its very condition. Only a sufficiently stable system can afford variations without collapsing. The field does not stifle the possible: it organizes it, making it habitable. It is a vision of life that is less mechanical and less romantic, but perhaps more realistic: life as an open equilibrium between constraint and freedom.

VI. Information Taking Shape.

If the previous chapter introduced the idea of the biological field as an organizing reality, this sixth chapter takes a further—and, in a certain sense, more delicate—step: it clarifies what we really mean by "information" when we speak of life. This is not a mere terminological detail. It is the point at which many scientific metaphors—powerful, yet often misleading—reveal their limitations.

In everyday language and much popular science writing, biological information is still described as a set of data, instructions, or codes: DNA as software, the cell as a computer, the organism as a programmed machine. This metaphor has been of great historical utility, but today it shows clear cracks. Data, on its own, does nothing. It can be copied, transmitted, or stored. Life, on the other hand, organizes itself, sustains itself, and transforms itself. This chapter aims to highlight this crucial difference.

The underlying thesis is simple only in appearance: biological information is not an object, but a relationship that takes on a st d form in matter. It does not govern life from the outside, like an abstract code, but emerges and operates from within vital processes. It is information because it makes a difference; it is embodied because it never exists separately from the structure that brings it into being.

Within the book's overall context, this chapter serves as a pivotal link. After demonstrating, in the preceding sections, the limits of biochemical reductionism and the importance of coherence and fields, this chapter clarifies what circulates, stabilizes, and transforms within those fields. Not indistinct energy, not digital instructions, but informational configurations that are simultaneously form, dynamics, and biological meaning.

Physiologist Denis Noble, one of the leading figures in the post-genomic debate, takes a critical stance toward the computational metaphor when he states unequivocally: *"There is no program in DNA*." From this perspective, the genome is not a self-contained instruction manual, but a resource that acquires meaning only within a network of cellular, tissue, and environmental interactions.

This chapter does not yet anticipate the clinical or cognitive implications that will be developed later. It deliberately limits itself to setting the record straight on an overused concept, showing why speaking of information in biology means speaking of active forms—organizations that persist over time and, precisely for this reason, can evolve, learn, and sometimes disintegrate.

Growing Without Instructions: The Embryo as a Self-Organizing Field.

If we truly want to understand what it means to say that life is information incarnate, there is no more revealing terrain than embryonic development. It is here, in the first hours and days after fertilization, that biology most clearly demonstrates a fact that is uncomfortable for any mechanistic view: form is not executed, but takes shape.

The embryo does not follow a detailed blueprint, like a house built from a technical drawing. Rather, it resembles a process that progressively finds its direction, stabilizing some possibilities and allowing others to fade away. Cells divide, migrate, and change function, but they do so without a command center telling them, "Now become a liver" or "Now become a retina." Yet, with surprising reliability, a coherent organism emerges.

As early as the 1930s, the German embryologist Hans Spemann, who won the Nobel Prize in 1935, demonstrated that certain regions of the embryo do not contain a pre-determined form but rather activate organizational processes in the surrounding tissue. They do not transmit detailed instructions; they create conditions. It is a subtle but decisive distinction. Information does not lie in the final result, but rather in the ability to guide a collective process.

From this perspective, embodied information is not something that "lies within" a cell like a sealed message. It is a relationship that extends across the space and time of the embryo, involving chemical gradients, mechanical tensions, division rhythms, and geometric constraints. Form emerges because the system, as a whole, recognizes and stabilizes configurations compatible with its own coherence.

The theoretical biologist Stuart Kauffman, reflecting on morphogenesis, expressed this point in words that resonate strongly with the idea of a field: *"Organisms are not built according to a blueprint; they self-organize within spaces of possibility."* Embryonic development thus becomes the natural laboratory in which to observe information as it becomes flesh, without ever separating from the matter that expresses it.

This perspective also helps us understand why the embryo possesses a remarkable regulatory capacity. If some cells are removed or damaged, the pro l process does not necessarily fail:

it often reorganizes itself. Not because the "program" is redundant, but because the information is distributed, embodied in the field of relationships. The final form is not a predetermined goal, but a dynamic attractor.

In this sense, embryonic development clarifies one of the central points of the chapter: biological information does not precede form as an abstract idea, nor does it follow it as a mere description. It coincides with the very process that generates form. The embryo does not apply information: it lives it.

Because life cannot be "restarted" like a digital system.

One of the most persistent illusions in contemporary biology stems from the—often implicit—comparison with digital technologies. A computer crashes; you reboot it. A file becomes corrupted; you restore it from a previous copy. But life doesn't work that way, and this chapter—at the heart of the reflection on embodied information—finds its most convincing proof in irreversible time.

A living organism does not pass through time as a neutral medium over which data flows. Time itself enters its structure, settling into its form, its tissues, and its developmental trajectories. Every moment lived alters the organism's informational field, leaving traces that cannot be erased without destroying the system that bears them. Biological information, precisely because it is embodied, is always also history.

This is the point at which the computational metaphor definitively breaks down. A digital system can revert to a previous state because information is separable from its medium. An organism, on the other hand, is its own information over time. There is no "before" preserved elsewhere from which to start anew. Even when a wound heals, one does not return to the initial state: one moves forward, with a new configuration.

As early as 1944, Erwin Schrödinger, in his famous *"What Is Life?"*, had grasped an essential aspect of this temporal asymmetry, speaking of life as a system that feeds on "negative entropy." Beyond the now-dated language, the insight remains powerful: *"A living organism maintains its order by increasing the disorder of its environment.*" This exchange is irreversible. Every act of maintaining form comes at a temporal cost that cannot be undone.

Biological information, therefore, is not merely a spatial relationship, but a temporal process that leaves traces. Consider the maturation of the nervous system, early imprinting, or even simply cellular aging. In all these cases, information is not "executed" and then stored: it becomes embedded in matter, forever altering its future possibilities. To "reset" would mean erasing history, but erasing history is equivalent to erasing the organism.

This is why life cannot be cloned in the strict sense of the term, not even when the DNA is identical. As the French biologist Henri Atlan observed, reflecting on complex systems: *"The information that matters is not the information that repeats itself, but the information that transforms itself.*" And transformation, by definition, happens only once.

In this light, temporal irreversibility is not a limitation of life, but the very condition that makes it possible. Without time flowing in a single direction, there would be no development, learning, or adaptation. Embodied information is always information that has occurred, not simply preserved. And precisely for this reason, every organism is unique: not because it contains special data, but because it has traversed a path through time that no other organism can ever retrace in the same way.

From physics to consciousness: the conceptual journey of information.

There are concepts that do not belong to a single discipline but function as epistemological bridges: they enable dialogue, reframe problems, and change the very way questions are posed. Information, in the context of field biology, is one such concept. Not because it explains everything—universal explanations are always suspect—but because it forces different sciences to grapple with what they truly mean by "reality."

In physics, information began to carry real weight when it was realized that it was not merely a way of describing systems, but something that constrained their behavior. In the late 1980s, in the hallways of the *Los Alamos National* Laboratory, there was a lively debate over whether information was a physical quantity in its own right or merely a convenient abstraction. Physicist Rolf Landauer, with a remark that has since become famous, put it in no uncertain terms: *"Information is physical."* He did not mean that information was a "thing," but that it does not exist without a medium, without a process, without a cost. This is a lesson that biology has often only partially grasped.

When we turn to living systems, in fact, information ceases to be neutral. It does not simply describe what happens; it participates in what happens. This is where epistemology comes into play. In biology, talking about information means asking who interprets it, how, and based on what constraints. A chemical signal, a spatial distribution, or a rhythmic variation are not information in and of themselves: they become information only within a system capable of giving them operational meaning.

This is the point of contact with the philosophy of mind. In the 1990s, the American philosopher John Searle liked to point out that manipulating symbols is not enough to produce meaning. Information, without a subject—or at least a system—to embody it, remains mute. Field biology takes this observation seriously but radicalizes it: every organism is already, in an embryonic form, a subject of meaning, because it is a center of informational integration.

An anecdote often recounted in European cybernetics circles concerns an informal meeting in Vienna between biologists and philosophers in the early 1970s. During a coffee break, someone asked whether the brain "stored" information like a library. Neurobiologist Walter Freeman, known for his studies on dynamic brain patterns, replied with a smile: *"No, the brain dances with information.*" An imperfect but illuminating metaphor: information is not a mental object; it is an event structured over time.

From this perspective, information becomes the meeting point of three traditions: physics, which views it as a constraint and a possibility; biology, which experiences it as a form stabilized within processes; and the philosophy of mind, which examines it as a condition of meaning.

It is no coincidence that thinkers as diverse as Gilbert Simondon, in France, and Thomas Metzinger, in Germany, have insisted—from opposing perspectives—that we cannot understand the mind without a theory of the processes that generate it. Information, understood not as content but as an operational relationship, allows us to avoid two pitfalls: material reductionism and immaterial mystery.

From an epistemological standpoint, this also means recognizing a shift in the status of scientific explanation. We no longer explain life by saying "what it is made of," but rather by explaining how it organizes itself and maintains its own internal meaning. Information does not close the discussion: it opens it. It functions as an imperfect but indispensable lingua franca among disciplines that would otherwise remain incommunicable.

Ultimately, field biology proposes a cultural—even more than a scientific—challenge: that information might be the concept through which we can collectively rethink nature, life, and mind, without reducing one to the other. Not a new dogma, but a shared space for inquiry.

Who is watching whom? The observer between the laboratory and lived experience.

All biology begins with a gaze. A microscope turned on, a notebook opened, a well-formulated question. But when the object of that gaze is a living organism, something unique happens: we find ourselves faced with systems that are not only observed but that, in their own way, observe themselves. It is here that the concept of embodied information reveals one of its most fruitful tensions: that between science's external observation and life's internal self-reference.

In the laboratory, the scientific observer operates through separation. They isolate variables, control conditions, and translate processes into measurements. It is a powerful approach that has yielded extraordinary results. But, as Claude Bernard already noted in mid-nineteenth-century Paris, *"life eludes us as soon as we break it down too much."* Bernard was not questioning the experiment itself, but rather pointing out that the organism reacts to the very act of being observed, because it is a system regulated from within.

Here a crucial distinction comes into play. Scientific observation is hetero-referential: the criterion of validity lies outside the observed system. Biological self-reference, on the other hand, is constitutive: an organism lives because it maintains a continuous relationship with itself, updating its internal information in the interest of survival. It does not "know" it is being observed, but it continuously observes itself, in the operational sense of the term.

In the 1960s, amid intense social tensions in Chile, biologists Humberto Maturana and Francisco Varela arrived at this insight while studying perception in nervous systems. Their concept of autopoiesis—officially presented in 1972—also arose from a startling empirical finding: the nervous system does not record the world as it is, but continuously produces internal coherence. As “ ” Maturana said in a lecture in Santiago, in a phrase that has since become famous: *“Everything that is said is said by an observer.”* But in living beings, that observer is already part of the system.

This capacity for self-reference had already been intuited, in a surprisingly modern form, by the Baltic biologist Jakob von Uexküll in the early 20th century. While studying simple animals, he introduced the concept of the Umwelt: every organism lives in a meaningful world that does not coincide with that of the human observer. There is no single biological reality, but rather a multiplicity of worlds centered on the organism.

From the perspective of field biology, this implies a subtle yet decisive epistemological consequence: the scientific observer describes information fields, while the organism inhabits and generates them. The two perspectives are not in conflict, but they are not superimposable. Confusing them leads to conceptual errors: either the organism is reduced to a reactive machine, or science is attributed a subjectivity that does not belong to it.

Life, as embodied information, is always self-centered, even in its simplest forms. A bacterium that regulates its own metabolism does not “know” in the human sense of the term, but it makes distinctions that matter to it. This is the lowest common denominator between biology and the mind: not reflective consciousness, but the ability to relate information to an operational identity. In this sense, the observer is not an intruder in the biological discourse, but a point of reference. Science observes life; life sustains itself by observing itself. Much of the difficulty—and the fascination—of a biology that truly seeks to go beyond reductionism lies between these two levels.

When information changes the way we know.

There was a moment, in the second half of the 20th century, when science began to use a seemingly innocuous word—information—to say something radically new about the world. It was no longer just a matter of measuring, weighing, and quantifying. It was about understanding how order is maintained, transmitted, and takes shape, even against the natural tendency toward disorder. Since then, epistemology—the very way in which science conceives of what it knows—has never been the same.

The story begins in an unexpected place: thermodynamics. In the 1940s, as Europe was struggling to emerge from the war, physicists such as Leo Szilard and Léon Brillouin began to ponder an age-old paradox: how can a system produce order without violating the second law of thermodynamics? The famous "Maxwell's demon," first conceived in the 19th century, was now being viewed in a new light. Szilard suggested that the act of knowing—distinguishing, measuring, selecting—carries an energy cost. Information thus entered physics not as a metaphor, but as an epistemic quantity.

A few years later, in 1948, Claude Shannon, working at Bell Labs in New Jersey, published *"A Mathematical Theory of Communication*." His goal was technical: to improve the efficiency of telecommunications. But the consequences were philosophical. By defining information in terms of uncertainty reduction, Shannon clearly separated information from meaning. This was a methodological choice, not an ontological one. Yet that separation would influence the way science spoke of knowledge, the mind, and life for decades to come.

During those same years, in Cambridge, Alan Turing was pondering what it means to "think." His famous test did not define intelligence based on subjective experience, but on the coherent manipulation of symbolic information. Thus was born a computational view of the mind that would dominate philosophy and the cognitive sciences for much of the second half of the century. Epistemology was shifting: knowing was no longer about mirroring the world, but about transforming information according to rules. And yet, something didn't add up. In 1944, in a small book written in Dublin during the war, Erwin Schrödinger posed a question that was destined to become famous: What is life? His answer was surprising: an organism is a system that feeds on "negative entropy." Today we would say: it maintains its form thanks to ordered information flows. Schrödinger did not speak of genes as codes, but of structural stability over time, opening a breach between physics, biology, and epistemology.

In the 1950s and 1960s, this breach widened with the advent of cybernetics. In New York, at the famous *Macy Conferences*, figures such as Norbert Wiener, Gregory Bateson, and Margaret Mead discussed feedback, regulation, and learning. It was here that epistemology shifted gears. It is no longer enough to ask whether a description is true or false. One must ask: For whom is it informative? In what context? With respect to what internal organization? These questions would also become central to the philosophy of mind in the 1970s and 1980s, when Hilary Putnam's functionalism began to show its limitations, paving the way for more embodied and situated conceptions of cognition. An anecdote recounted by a Dutch philosopher of science—who in the 1980s frequented both neurobiology laboratories and seminars in analytic philosophy—captures the atmosphere of the time well. "We were all talking about information," he recalled, *"but we were no longer talking about the same thing*." For physicists, it was a measure; for biologists, an organizational function; for philosophers, a problem of meaning. It was precisely this ambiguity that made information an epistemologically fruitful concept.

In the context of field biology, this history suggests a clear conclusion: information is not a neutral intermediary between subject and object, but a principle that redefines both. We no longer know a world made up of things, but of informational processes that stabilize into forms. Life, in this framework, does not “contain” information: it is information that has found a way to endure in matter.

Because life does not function like a computer.

There is an image that profoundly shaped biology in the second half of the 20th century: that of the organism as a machine that processes information, of DNA as software, and of cells as sophisticated hardware. A powerful, seductive metaphor—and one that was extremely productive on a technical level. But, as often happens, what initially helps us understand risks, over time, becoming a distorting filter.

The computer metaphor emerged in a specific historical context. It was the 1950s and 1960s, when the first computers occupied entire rooms—such as the IBM 7090 installed at MIT—and cybernetics promised a common language for machines, brains, and society. Speaking of "information" in biology at that time meant stepping into the future, breaking free from vitalism and philosophical vagueness. But the price paid was high: information was progressively reduced to data, to a discrete sequence, to an instruction to be executed.

The problem is that data do not live. They are abstract, interchangeable entities, identical to themselves regardless of context. A file remains the same on a computer in Tokyo or Turin. An organism, on the other hand, does not. The same "information" produces different effects depending on the state of the system receiving it, its history, and its environment. Confusing information with data means ignoring this fundamental dimension: dependence on context and form.

In the 1990s, the British physiologist Denis Noble, while working at Oxford on mathematical models of the heart, often recounted a curious episode. During a conference, a computer scientist asked him, *"Where is the program for the heartbeat?"* Noble paused awkwardly before replying, *"There is no program. The heart beats because it is built a certain way."* Not to execute code, but because of a distributed dynamic involving cells, tissues, electric fields, and mechanical constraints. Information, in this case, is not written anywhere: it emerges from the organization itself.

The computer science metaphor, when applied to life, also introduces another misconception: the idea that information precedes matter and governs it from the outside. As if DNA were a text to be read and the body a biological printer. But this view does not hold up to scrutiny. As early as the 1970s, studies in experimental embryology—such as those conducted in Naples by Giuseppe Levi and his school—showed that the form of an organism depends on spatial and temporal interactions, not on a simple linear execution of instructions.

A piece of data, by definition, is passive. Biological information, on the other hand, is active and performative. It does not describe what happens: it makes it happen. This is where the computer metaphor breaks down, because in computers, meaning is irrelevant. A sequence of bits functions independently of what it "means." Not so in organisms. As the Chilean biologist Humberto Maturana observed while speaking in Santiago in the 1980s, *"what matters is not the signal itself, but how the system responds."*

An anecdote collected by a French biologist who worked in the 2000s at a genomics laboratory in Lyon is illuminating. She recounts how two cells with the same genetic makeup reacted oppositely to the same chemical stimulus. "*At first we thought it was a mistake*," she said, *"then we realized that we were treating information as data, and the cells were telling us that it wasn't."* There was no bug in the system: there was a living organism that was interpreting it.

Criticizing the computing metaphor does not mean rejecting technology or computational models. It means putting them in their proper place. A model can simulate, describe, or approximate. But life does not "process" information the way a machine does: it embodies it, transforming it into structure, gesture, and rhythm. Reducing everything to data means losing what makes an organism alive: its capacity to make sense of things, to maintain coherence, and to respond creatively.

From this perspective, field biology proposes a shift in perspective. Information is not a sequence to be decoded, but a relationship that takes shape. Not a message that travels, but a configuration that stabilizes. Abandoning the computer metaphor is not a step backward, but an act of epistemological maturity: recognizing that life is not a misunderstood computer, but a phenomenon that demands new concepts.

Because an algorithm doesn't know it's alive.

Every era conceives of life using the tools at its disposal. In the seventeenth century it was the clock, in the twentieth century the thermodynamic machine, and today it is the computer. It is therefore not surprising that, in the age of artificial intelligence, the organism is often described as a particularly sophisticated computational system. But the comparison, however evocative, soon reveals its flaws. Because the information that a machine computes is not the same information that an organism experiences.

In 1956, during the famous *"Dartmouth Summer Research Project on Artificial Intelligence*," John McCarthy and Marvin Minsky envisioned machines capable, in principle, of doing everything a human mind does. The optimism was contagious. In those same decades, in molecular biology laboratories, people spoke of "reading," "transcribing," and "translating" the genetic code. The living world finally seemed decipherable like a formal language.

Yet, seventy years later, a comparison between artificial intelligence and biological intelligence reveals a radical difference that goes beyond computational power. AI manipulates symbols, even when it does so with surprising effectiveness. The organism, on the other hand, is intimately involved in what it knows. For a living being, information is never neutral: it has vital significance, guides action, and shapes form.

An anecdote recounted by an Italian researcher who worked for many years between Milan and Boston in the field of *"machine* learning" is emblematic. During a computer vision test, a system perfectly recognized a human face even when the image was distorted or incomplete. *"Better than us,"* someone commented. The researcher replied, *"Yes, but it doesn't know what it means to meet that gaze."* The machine classified; the human being entered into a relationship.

That is the point. Computed information is external to the system that processes it: it can be transferred, copied, or deleted without ontological consequences. An algorithm does not suffer if it loses data. Biological information, on the other hand, is intrinsic: it coincides with the very state of the organism. When an animal perceives danger, the information does not simply "arrive": it translates into posture, muscle tension, and hormonal changes. It is information that takes on a physical form.

As early as the 1980s, neuroscientist Gerald Edelman, winner of the 1972 Nobel Prize in Medicine, emphasized this point during his lectures at La Jolla: the brain does not function like a computer because it never separates processing from value. Every cognitive act is selective, situated, and tied to the organism's individual history. Artificial intelligence can optimize a function; a living being must continue to exist.

Even when AI mimics adaptive behaviors—as in the *"reinforcement* learning" systems developed since the 2010s—a decisive difference remains. A machine's "reward" is a numerical parameter. For an organism, what is at stake is the continuity of its own living form. Here, information is not just one input among others: it is what maintains the coherence of the biological field.

A little-known but significant episode dates back to an interdisciplinary workshop held in Santa Fe in the early 2000s. An evolutionary biologist observed that if a robot falls, it gets back up because it is programmed to do so; if a child falls, it gets back up because it has learned what it means to fall. It is not merely a difference in complexity, but in the nature of the information involved. In the first case, it is calculation; in the second, it is experience embedded in the body.

Field biology offers a decisive interpretive key here. Biological intelligence does not reside in a control center, nor in a hidden algorithm, but in the overall coherence of the system. Information is the way in which the field organizes itself and sustains itself over time. For this reason, it is inseparable from matter, energy, and form. A machine can simulate intelligent behavior; an organism is that behavior, in the full and risky sense of the term.

Recognizing this difference does not mean belittling artificial intelligence, but rather freeing it from inappropriate expectations. And, above all, it means restoring life's specificity: not as biological hardware that runs evolutionary software, but as a process of lived, embodied, irreversible information. A subtle yet decisive difference that marks the boundary between what calculates and what lives.

From the bug to change: two opposing fates of error.

In the language of computer science, error is an intruder—a bug to be identified, isolated, and corrected. In biology, the exact opposite is true: error is not an enemy of life, but one of its deepest driving forces. This seemingly subtle difference exemplifies what means when it states that life is embodied information and not merely data processing.

When Grace Hopper's team found a moth stuck in a relay of the Mark II computer at Harvard in 1947, the incident—now preserved at the Smithsonian—became the founding anecdote behind the term *"bug."* The error was something external, accidental, to be removed in order to restore the machine's proper functioning. Once the cause was eliminated, the system returned to exactly how it was before. No trace, no memory, no transformation.

In living organisms, the opposite occurs. A genetic error, a variation in development, or a "wrong" physiological response is not simply erased: it becomes inscribed in the system's history. It can be reabsorbed, compensated for, or open up unexpected paths. In 1943, Conrad Waddington observed in his experiments at Cambridge how embryos subjected to thermal stress developed new morphological configurations, some of which stabilized over time. It was not a malfunction to be corrected, but a deviation that redefined the range of possibilities.

Here a crucial difference emerges between computed information and lived information. In a machine, an error is a discrepancy between the expected state and the actual state. In an organism, an error is an event—something that happens to the system and within the system, altering its internal coherence. Life never simply returns to its previous state: it incorporates what has happened.

A French physician recounted how a serious loading error during a patient's rehabilitation—a premature movement, theoretically against the rules—had triggered an unexpected reorganization of posture. The body had "made a mistake," but that mistake had forced the nervous system to find a more stable solution. In a control algorithm, the same error would have triggered an alarm; in an organism, it had generated learning.

This is because, in field biology, information is not a message that passes intact from one point to another, but a relationship that restructures itself. Error does not interrupt the flow of information: it diverts it, tests it, and sometimes enriches it. As Georges Canguilhem wrote in **The Normal and the* Pathological* (1966), life is not passive adaptation to a norm, but the capacity to establish new ones. And often it is precisely error that makes this invention necessary.

Even in the most advanced artificial intelligence technologies, error remains fundamentally reversible. A model is retrained, the weights are updated, and the loss function is minimized. Nothing that has happened "leaves scars." In living organisms, however, every significant error leaves a mark: an epigenetic trace, an immune memory, a behavioral change. The living being does not simply correct itself: it transforms itself.

In this sense, biological error is not the opposite of information, but one of its generative modes. It is the moment when the field temporarily loses coherence and, precisely for this reason, is forced to reorganize itself at a more complex level. Machines seek functional perfection; life seeks continuity. And continuity, paradoxically, often passes through error.

The Incomplete Code: Epigenetics, Space, and Time in Field Biology.

For over half a century, we have told the story of life through letters: A, T, C, G. An elegant alphabet, etched into the double helix of DNA, capable—so it was said—of explaining the shape of a leaf, the beating of the heart, even a personality trait. It is a powerful narrative, born in the laboratories of Cambridge in the 1950s and quickly becoming a founding myth of modern biology. But as is often the case with successful myths, over time it began to show cracks.

As early as the 1970s, in Woods Hole, during the famous *summer schools* attended by biologists, physicists, and philosophers of science, a joke attributed to an elderly researcher was making the rounds: *"DNA is like a list of ingredients. But where are the recipe, the kitchen, and the cook?"* It was a provocation, of course. But it foreshadowed a real problem: the genome, on its own, does not explain how information becomes an organism.

The idea that DNA contains a complete "program" is now difficult to sustain. Not because genes don't matter—they certainly do—but because their meaning depends on context. A gene can be silenced, amplified, or modulated. It can be active in one tissue and remain silent in another. The same genetic sequence, under different conditions, produces different outcomes.

This is where epigenetics comes into play, a term coined as early as 1942 by Conrad Waddington in Edinburgh, long before it became fashionable. More recent experiments have made his intuition almost self-evident. In the 2000s, studies of monozygotic twins—who are genetically identical—showed that, al , their epigenetic profiles diverge over time. DNA methylation, histone modifications, microRNAs: small chemical marks that do not change the letters of the code, but alter how it is read. Life, we might say, annotates the margins of the genetic text.

There is also an even more radical aspect: biological information is spatial. It resides not only in the sequence but in the arrangement. An embryo does not grow like a 3D printer executing linear instructions; it grows as a field of relationships. Cells "know" where they are, who is next to them, and what shape is emerging.

In the 1950s, the British biologist Lewis Wolpert proposed the concept of *"positional information*": cells read chemical gradients and, based on their position, activate different programs. A simple idea, born from observing insect development, which nevertheless undermined the notion of centralized control. Form is not imposed from above; it emerges from a distributed map.

An anecdote often told in the halls of the University of Cambridge concerns a teaching experiment: students are asked how a cell in the hand knows it is a "hand" and not a "liver." The instinctive answer is, *"Because the DNA says so.*" But the DNA of a cell in the hand and a cell in the liver is the same. The difference lies in space, in interactions, in local history. Here, information is topological.

Further complicating the picture is the dynamic dimension. Biological information is not static: it unfolds over time. Oscillations, rhythms, cycles. From the pulsation of the circadian clock genes—discovered in the 1990s while studying Drosophila—to the waves of electrical activity in developing tissues, life organizes itself as a temporal process.

A fascinating example comes from regeneration. In the 1960s, in laboratories in Chicago, researchers observed how planarians could regenerate an entire organism from a fragment.

It is not enough to say that "DNA knows what to do." The real question is: how does the fragment know whether to regenerate a head or a tail? More recent studies suggest that bioelectric signals and dynamic membrane patterns play a crucial role. It is information that flows, that stabilizes, and that can also be disrupted. Here, field biology finds one of its strengths: the organism as a stabilized informational structure, maintained by continuous flows. Not an object, but an event that endures.

To say that life goes "beyond DNA" does not mean abandoning science in favor of metaphor. It means recognizing that biological information is embodied: it takes shape in space, in time, in organized matter. It is not a file to be read, but a relationship to be maintained.

In 2001, upon the publication of the first draft of the human genome, some spoke of the "book of life." Years later, one of the project's leading figures, in an interview with The New York Times, admitted with a touch of irony: *"We have the dictionary, but not the grammar."* Perhaps today we can add: nor the prosody, the rhythm, or the arrangement of sentences on the page.

Field biology proposes precisely this shift in perspective. Information not as a command, but as an active configuration. Not as a linear cause, but as a coherence that sustains itself. From this perspective, DNA remains fundamental, but ceases to be the absolute ruler. It becomes an instrument within a larger orchestra, where space, time, and relationships make the music.

Memory without an archive: how cells remember who they are.

When we speak of memory, our thoughts immediately turn to the brain, to childhood memories, to a yellowed photograph. Yet memory is much older than the mind. It existed before neurons, before nervous systems, perhaps even before anyone could give it a name. Field biology invites us to view life as a , information-rich structure stabilized over time, and within this framework, non-genetic memory emerges as one of its most revealing phenomena.

Cells remember. Not in a metaphorical sense, but in a functional way: they retain traces of past states that influence their future behavior, without any modification to their DNA. It is an embodied, distributed, and often silent memory—a memory that does not need to be written down to take effect.

In the 1980s, in the Pasteur Institute laboratories in Paris, researchers studying cell cultures noticed a curious phenomenon: cells subjected to temporary stress—a temperature change, a nutritional deficiency—reacted differently to a similar stress applied days later, even when initial conditions had returned to normal. There was no detectable genetic mutation, yet something had remained. A sort of cellular "experience."

Today we know that this memory can reside in many places: in the conformation of proteins that remain active or inactive, in the organization of the cytoskeleton, in the composition of the extracellular matrix, and even in the biophysical patterns that run through tissues. A cell never truly starts from scratch. Every state leaves an imprint.

A fascinating example comes from cell differentiation. When a stem cell becomes, say, a muscle cell, it enters a stable state that tends to persist. Even when removed from its original context, it often "remembers" that it is muscle. This memory of identity is not solely genetic: it is an overall configuration involving form, mechanical tensions, and protein networks. Field biology would describe this as a coherence that has stabilized.

There is a little-known but instructive chapter in this story concerning *prions*. Discovered in the 1980s by Stanley Prusiner—who received the Nobel Prize in 1997 for this work—prions are proteins capable of existing in multiple stable conformations. What is surprising is that a protein can "teach" its own shape to other identical proteins, without involving a gene.

In yeast, some prion conformations are not pathological but functional: they act as a molecular memory that alters cellular behavior for many generations. It is a radical example of information without a code: the form itself is memory.

Non-genetic memory becomes even more evident at the tissue level. Those working in reconstructive surgery know this well: scar tissue "behaves" differently from healthy tissue, even years later. It is not just a matter of different cells, but of a mechanical and structural history imprinted on living matter. In the 1990s, in Boston, several studies on tissue engineering showed that cells cultured on rigid or elastic substrates developed different fates. But the most interesting discovery came later: even when transferred to a new substrate, the cells retained the "memory" of their previous rigidity for some time. It was as if the mechanical past continued to speak.

This type of memory defies the computer science metaphor. There is no file saved somewhere. Rather, there is an embedded history, similar to the trail left by a path through the woods: the more it is walked, the more evident it becomes.

This idea of diffuse memory has profound philosophical implications. Henri Bergson, in his lectures at the Collège de France, distinguished between memory as a "repository" and memory as "duration." For Bergson, life does not preserve the past like an archive, but carries it along as a continuous momentum. In biology, non-genetic memory seems to give substance to this insight. Paul Ricoeur, too, many decades later, when speaking of personal identity, emphasized that remembering does not mean reproducing the past, but maintaining a trace of it that is active in the present. Cells, in their own way, do something similar: they do not "remember" what has been, but incorporate it into what they are.

From this perspective, the organism appears as a field of layered memories. Each level—molecular, cellular, and tissue—preserves traces that guide the future. Memory is not an addition to life; it is one of its conditions.

Seen this way, remembering does not mean remaining the same, but remaining consistent. Field biology describes the organism as a stabilized information structure, and non-genetic memory is one of the mechanisms that make this stability amid change possible.

An Italian biologist, speaking informally at a conference in Padua in the early 2000s, summed up the concept with a phrase that has remained famous among students: *"Cells don't know where they come from, but they don't forget it.*" It's a joke, of course. But it captures an essential point: life does not proceed in isolated moments. Every moment is imbued with what preceded it.

Aging as an accumulation of biological memories.

There is a reassuring idea that has shaped our understanding of biology for decades: if DNA remains intact, the cell's identity is preserved. Yet aging undermines this certainty. Genetically identical organisms—think of monozygotic twins—can age in surprisingly different ways. Field biology offers a powerful interpretive framework: cells do not age merely because they break down, but because they remember. And they do not always remember well.

From this perspective, the organism is a stabilized information structure that spans time. Aging is not merely wear and tear, but the accumulation of past states that remain operational: functional memories that enable adaptation and dysfunctional memories that, over time, stiffen the system until senescence sets in.

In 2013, at a conference on longevity held at Cold Spring Harbor—one of the iconic sites of molecular biology—a researcher presented a series of images of cell nuclei taken from young and aged tissues. The differences lay not in the DNA sequence, but in the architecture: fragmented heterochromatin, a less stable nuclear lamina, and blurred boundaries. It was like looking at a library where the books are the same, but the shelves have warped.

This loss of structural order is a form of memory. Every cell division, every inflammatory stress, every metabolic imbalance leaves a trace in the cell's internal organization. Over time, these traces accumulate. The cell "knows" it is old because maintaining its informational coherence has become more taxing.

A striking example comes from the nuclear lamina, the network of proteins that supports the nucleus. In progeria—a rare disease of premature aging first described in 1886 by Jonathan Hutchinson in London—the DNA is nearly normal, but the structure that organizes it is compromised. The result is accelerated aging: a dramatic demonstration that the memory of age can be architectural, not genetic.

Many cellular memories originate as survival strategies. A cell that has faced repeated oxidative stress learns to defend itself: it activates response programs, modifies its metabolism, and slows proliferation. In the short term, this is a wise choice. In the long term, it can become a death sentence.

This is what happens in replicative senescence, first observed in the 1960s by Leonard Hayflick in Philadelphia. Cells stop dividing not because their DNA is irreparably damaged, but because they have "decided" to stop. This decision is a memory: the recollection of a past risk that continues to influence the present.

The problem is that senescent cells do not remain silent. They secrete inflammatory signals that alter the biological environment of the tissue. It's a domino effect: one cell's memory becomes the environment for many others. Aging, then, is no —it is not the sum of old cells—but the propagation of dysfunctional memories within a shared environment.

One of the most talked-about experiments of the past twenty years originated at the University of California, Berkeley. In

2005, Irina and Michael Conboy studied heterochronic parabiosis: they connected the circulatory systems of young and old mice. The tissues of the older mice showed signs of rejuvenation. The DNA had not changed. What had changed was the informational context.

The most intriguing interpretation, from the perspective of field biology, is that the older cells had not "forgotten" how to function, but were trapped in an impoverished field. When placed in an environment rich in coherent signals, some age-related memories became reversible. Not all of them, of course. But enough to suggest that aging is not a one-way street. During an informal visit to the Berkeley lab, a regenerative biologist made a comment that has remained in many students' notes: *"We don't have old cells; we have cells living in an old world."* It's a provocative statement, but it hits on a crucial point.

In recent years, there has been much talk of epigenetic clocks, such as the one proposed by Steve Horvath in 2013. They measure biological age through DNA methylation patterns. But more than clocks, they are memory recorders: they read how the organism's life history has been imprinted on its regulatory systems.

A person who has lived for a long time under conditions of chronic stress may show a biological age higher than their chronological age. Not because their genes have changed, but because the organism's information field has incorporated that experience. Age, in this sense, is not just the passage of time, but the way in which it is retained.

Here, a dialogue with culture is inevitable. Marcel Proust wrote that true lost time is that which leaves no trace. In biology, the opposite occurs: time that leaves too many traces can become a burden. Aging is a memory that has ceased to be plastic. If life is embodied information , then aging is a narrative crisis. The organism struggles to integrate its past into a coherent form. Cellular memories are no longer reworked but accumulate like chapters that have not been rewritten.

In this light, senescence appears less as a failure and more as a rigidity: the difficulty of updating one's biological narrative. Field biology does not promise eternal youth, but it offers a shift in perspective: understanding age as an emergent property of the organism's distributed memory.

The immune system does not forget...

For a long time, we have described the immune system as a well-organized library: shelves of antibodies, each labeled with the name of a pathogen it has encountered. A convenient, effective metaphor, but one that is increasingly inadequate. Today we know that immunity is not limited to storing molecular "copies" of the past. Rather, it remembers by changing itself.

From the perspective of field biology, immune memory is not a saved file, but a persistent reorganization of the organism's cellular field. After an infection, it is not just specific antibodies or lymphocytes that remain: what remains is a new way of being for the immune system, a configuration that guides future responses.

In 1957, Frank Macfarlane Burnet, working in Melbourne, formulated the theory of clonal selection. It was a breakthrough: immunity does not create antibodies "on demand," but selects cells that are already predisposed. For this work, Burnet received the Nobel Prize in 1960. But, as often happens, a brilliant theory ended up becoming an interpretive cage.

Clonal selection provides a good explanation for specificity, but less so for the systemic duration of memory. Why, after an infection, do even cells that have never directly encountered the antigen undergo changes? Why do tissues react differently—more rapidly, more coordinately? The answer that has emerged in recent years is clear: immune memory is distributed. It involves the bone marrow, lymph nodes, and peripheral tissues, but also cellular metabolism, the baseline inflammatory state, and communication between immune and non-immune cells. It is the entire field that is being reconfigured.

For decades, it was widely accepted that only adaptive immunity possesses memory. Then, in the early 2000s, this certainty began to crumble. In Nijmegen, the Netherlands, Mihai Netea and his team observed that certain innate immune cells—such as monocytes—responded more strongly to subsequent infections, even those different from the first. Thus was born the concept of *"trained immunity."* It had nothing to do with antibodies or specific receptors, but rather with a stable change in cellular behavior: altered metabolism, more accessible chromatin, and different activation thresholds. In other words: a memory without an archive, a competence embedded in the way the cell functions.

During a seminar in Bonn in 2016, a clinician described elderly patients who responded better to infections after receiving a "non-specific" vaccine such as BCG. It wasn't magic, but a reorganization of the immune system. Once "trained," the system remained more vigilant.

Another key piece of the puzzle comes from the discovery of tissue-resident T cells (Trm). Unlike lymphocytes that circulate in the blood, these cells remain in the sites where the infection occurred: the skin, lungs, and intestines. They were studied extensively following the flu epidemics of the early 2000s, particularly in laboratories in La Jolla, California.

Tm cells do not merely retain antigenic specificity; they retain a memory of the location. They are shaped by the tissue microenvironment, by local signals, and by the inflammatory history of that site. When the pathogen returns, the response is not only faster—it is more contextual, more "situated." Here, the biology of fields finds a concrete example: immune memory is not abstract, but spatial. The body remembers where it has been injured.

Then there is a silent protagonist: the microbiota. Studies conducted since the 2010s, jointly by Harvard and INSERM in

Paris, show that the gut microbial balance profoundly shapes the immune system. It does not merely stimulate it; it educates it. A gut that has coexisted for a long time with a certain microbial community develops active tolerance—a memory of coexistence. When this community is disrupted—by antibiotics, extreme diets, or prolonged stress—the immune field loses its coherence. Reactions become excessive or ineffective. Not because information is lacking, but because the relational memory has been disrupted. A gastroenterologist from Milan recounted, during an informal meeting, that patients "no longer reacted as they used to" after repeated courses of antibiotics. It was not a loss of defenses, but a loss of shared history.

Even the success of vaccines, viewed through this lens, appears in a different light. Vaccination does not merely mean inducing antibodies, but imprinting a new configuration on the immune system. Some vaccines work better than others not because of the magnitude of the response, but because of the quality of the reorganization they induce. This was clearly seen during the COVID-19 pandemic, when it was observed that protection did not always correlate with antibody titers. At play were tissue memories, trained innate responses, and underlying inflammatory states. The immune system, once again, revealed itself to be a field that retains its shape, not a biological hard drive.

Viewed in this light, immune memory is an act of coherence over time. The system does not preserve the past out of nostalgia, but to remain true to itself in the face of change. When this coherence becomes rigid, autoimmune diseases arise. When it dissolves, infections increase. In both cases, the problem is not a lack of information, but its poor integration. The immune system, as Elias Canetti wrote when discussing the masses, is "memory in motion." It does not accumulate; it transforms. And in this continuous transformation, it perhaps embodies one of the clearest examples of what life means as information taking shape.

The memory of living beings, from the cytoplasm to the forests.

There is a seemingly simple idea that unites a cell and a forest: neither reacts to the present as if it were a blank slate. Every response, every adaptation, every crisis carries a history with it. In field biology, this history is not a hidden archive, but a stabilized informational structure that guides what is possible and what is no longer so.

In ecology, there is increasing talk of *"legacy* effects." This refers to the way in which the past—fires, droughts, cultivation, extinct species—continues to influence the present. Not as a conscious memory, but as a configuration of the ecological field that makes some outcomes probable and others nearly unattainable. Exactly as happens in cellular memory.

In the 1980s, Canadian biologist C.S. Holling introduced the concept of ecological resilience while studying boreal forests and defoliating insect populations. His insight went against conventional wisdom: an ecosystem is not healthy because it is stable, but because it is capable of absorbing disturbances without losing its functional identity.

But Holling observed something more subtle. Two seemingly similar ecosystems, when subjected to the same disturbance, reacted in opposite ways. One regenerated; the other collapsed. The difference lay not in the event itself, but in ecological memory: in the sequence of previous states that had shaped the system.

A lake that has already experienced episodes of eutrophication, for example, can more easily "slip back" into a degraded state. The sediments, laden with phosphorus accumulated over the years, function as a chemical and biological memory that persists even when surface conditions improve. The lake remembers.

In the 1990s, studies conducted in the Caribbean Sea showed that many coral reefs, following bleaching events and overfishing, never returned to their original state. Filamentous algae grew in place of the corals. This was not a transitional phase: it was a new, stable regime.

Ecologists referred to these as "stable states." In terms of field biology, we could say that the ecosystem field had changed form. The memory of previous disturbances—loss of herbivores, increased nutrients, thermal stress—had reorganized the fundamental relationships. Returning to the previous state was not impossible in theory, but unlikely in practice.

The same phenomenon is observed in deserts that expand following intensive grazing. Once vegetation cover and soil structure are lost, the ecosystem enters a state that "remembers" aridity even when rainfall returns. The compacted soil, depleted of microbes and seeds, is an embodied memory of degradation.

Few places illustrate this idea better than the soil. A teaspoon of fertile soil contains billions of organisms, but above all, it contains stories of use, cultivation, fertilization, and abandonment. Agronomists and soil scientists know this well: a field does not respond immediately to a change in management.

In the 2000s, studies conducted in the countryside of eastern Germany, on formerly collectivized land, showed that even after twenty years of organic farming, some soils retained the "imprints" of the previous intensive chemical farming. It wasn't just a matter of residues: the soil's structure and micro , the dynamics of humus, and its ability to retain water had all changed. The soil remembered its industrial past.

Here, the parallel with cellular memory is evident. Just as a cell that has undergone repeated stress permanently alters its functional structure, so too does an ecosystem bear the imprint of the experiences that have shaped it.

A particularly telling example comes from the Mediterranean and California forests. For centuries, low-intensity fires were part of their life cycle. Then, in the 20th century, a policy of total fire suppression altered this dynamic. When fires returned—inevitably—they did so with devastating force. Forests "that had forgotten fire" burned with unprecedented violence. Ecologists today speak of *"fire memory"*: the frequency and type of past fires determine the ecosystem's future response.

In 2018, after the Camp Fire in California, a ranger reported that some areas showed no signs of spontaneous regeneration. Not because seeds were lacking, but because the soil and vegetation had lost their memory of the natural fire cycle. The forest ecosystem had become disorganized.

Not all memory is an advantage. Just as dysfunctional memories exist in cells, so too can the past become a burden in ecosystems. Overfishing, fragmentation, and pollution leave traces that reduce the capacity for adaptation. The collapse of the cod fishery in Newfoundland in the early 1990s is often cited as an example. Despite the near-total ban on fishing, the populations never returned to their previous levels. The system had crossed a threshold: new species had occupied the vacant niches, and the food web had reconfigured itself. The ecosystem "remembered" the absence of cod. In terms of field biology, we could say that the system had lost part of its historical coherence.

What emerges when we look at ecosystems is a powerful lesson: information is not contained within the elements themselves, but in the relationships that endure over time. Ecological memory cannot be localized in a single organism, but lies in the way flows of energy, matter, and signals have stabilized.

As in the cell, so too in the ecosystem, remembering means being oriented—not toward the past, but toward a future limited by what has been. Life, at every scale, does not simply react: it responds based on its own embodied history.

Stones That Remember: When Matter Preserves Its Own History.

Anyone who has run their fingers over the marble of an ancient sculpture or closely observed a restored fresco knows that matter is never neutral. It bears marks, scars, and layers. It does not merely "exist" in space: it tells a story. It is precisely in this almost tactile evidence that art and architecture offer a powerful cultural metaphor for biological memory—understood not as an archive of data, but as information that takes shape and persists in matter.

In the language of field biology, an organism is a stabilized informational structure. In art, we could say something similar: a form is never the instantaneous result of an idea, but the outcome of a long and often conflictual dialogue between design, material, time, and events.

Michelangelo knew this well. When he spoke of "freeing the figure" from the block of marble, he was not using just any poetic metaphor. Carrara marble, already quarried by the Romans, exhibited veins, microfractures, and internal tensions. It was not an indifferent material. Art historians note that, while working on the Moses (begun in 1513), Michelangelo changed his design several times in response to the limitations imposed by the block. The material "remembered" its own geological and technical history, and that information shaped the final form. This is not unlike what happens in a living organism, where structural history determines what can emerge.

Walking through Rome, Istanbul, or Jerusalem, one continually comes across buildings that are veritable material palimpsests. Churches built atop temples, mosques incorporating basilicas, medieval walls grafted onto Roman foundations. Architecture does not erase: it layers.

A prime example is Santa Maria sopra Minerva in Rome. The name itself is a declaration of memory: a Gothic church resting on an ancient temple dedicated to the goddess Minerva. The foundations are not merely a physical support but a persistence of information that spans the centuries.

In biology, we would say that the current form cannot be understood without the field's previous configuration. The same is true in architecture: every intervention engages in a dialogue with an embedded memory that resists the pure will of the designer.

In the twentieth century, the debate on restoration made this insight explicit. Cesare Brandi, in his **Theory of* Restoration* (1963), insisted on a crucial point: restoration does not mean returning a work to a hypothetical original state, but rather recognizing and respecting its material history.

A restorer recounted, during work on Giotto's frescoes in Assisi following the 1997 earthquake, that certain cracks should not be "corrected" but understood. They were the result of centuries of settling, vibrations, and previous interventions. Erasing them would have meant losing information. The parallel with biology is clear: intervening on an organism without taking its structural memory into account is equivalent to ignoring the field that holds it together.

There is a term dear to restorers and art historians: patina. It is not dirt; it is not decay. It is the mark of time that stabilizes a form, making it what it is. Walter Benjamin, speaking of the work of art, noted that aura also arises from this temporal irreproducibility.

Patina is information incarnate: it tells the story of exposure to light, air, and human contact. Consider the worn steps of medieval cathedrals, worn down by the passage of millions of feet. That form was not designed; it emerged. It is memory made visible. Similarly, in field biology, a living form is not merely the execution of a code, but the result of a history of interactions that leaves lasting traces.

A Japanese architect, Tadao Ando, explained in an interview in the 1990s how reinforced concrete—a seemingly anonymous material—reacts differently depending on humidity, temperature, and curing time. *"You don't build solely with the design,"* he said, *"you build with the memory of the material."* This resistance is not a flaw: it is a form of dialogue. As in living organisms, matter is not a passive carrier of information, but an active participant in its stabilization.

Art, architecture, and restoration teach us that memory is not something added later. It is what makes form possible. Matter that remembers is not nostalgia: it is structure.

From this perspective, life appears less like a machine executing instructions and more like a work in constant rewriting, where every present state is inseparable from its own embedded history. Just like a sculpture, a building, or a fresco that continues to live because it has never forgotten what it once was.

Philosophical Implications of Biological Space.

If biological information depends on space, then one consequence becomes inevitable: life is never abstract. It does not exist as a general formula, as an algorithm that can be replicated everywhere in the same way. Life always happens somewhere, and that "somewhere" is not a mere, ne l container, but an active component of the life process. This seemingly simple idea has profound philosophical implications, capable of challenging some of the most deeply ingrained habits of modern scientific thought.

For centuries, science has sought universal laws by separating them from their context. This worked well for planets and ideal gases. But when we enter the realm of the living, space ceases to be a backdrop and becomes a co-author. Jakob von Uexküll, an Estonian-German biologist and philosopher, had already intuited this in the early decades of the 20th century when discussing the Umwelt: every organism lives in its own world, structured by its perceptual and operational possibilities.

Transposed within the organism, this insight becomes even more radical. If information takes shape through spatial relationships—between molecules, structures, and fields of interaction—then we can no longer think of life as something that exists in space, but as something that produces space. Every cell, every tissue, every organism constructs its own internal space, maintains it, defends it, and reorganizes it. Field biology suggests that to live literally means to hold a coherent space together.

This perspective brings biology closer to a philosophical tradition often marginalized in scientific discourse. Maurice Merleau-Ponty spoke of the body not as an object in space, but as the origin of lived space. The body, he wrote, is not "in" a place: it is what makes a place possible. Applied to biology, this idea reverses the perspective: the organism does not occupy an environment; it structures it.

The implications are not merely theoretical. If life is inseparable from its place, then scientific experimentation, too, takes on a new status. An anecdote recounted by a French cell biologist during a seminar in Lyon in 2018 is illuminating. While studying cancer cells cultured in vitro, she noticed that certain "aggressive" behaviors disappeared when the same cells were reintroduced into a three-dimensional tissue more similar to their original environment. *"It wasn't th r the cells that had changed,"* she said, *"it was their place that had changed."* Space, in that case, did not explain life: it transformed it.

This gives rise to a new conceptual responsibility. Separating an organism from its context—whether it be an ecosystem, a tissue, or a microenvironmental niche—is never a harmless operation. It means altering the very information that organism embodies. In this sense, field biology engages in dialogue with deep ecology, but also with the ethics of science: if place is part of life, then intervening in place is equivalent to intervening in biological identity.

It is no coincidence that some of today's crises—from biodiversity loss to environment-related chronic diseases—can be interpreted as crises of dislocation. Organisms torn from their spaces of meaning, forced to function in fields they no longer recognize. Life, deprived of its place, does not immediately cease to exist, but it loses coherence. And the loss of coherence, as we will see in the following chapters, is often the first step toward disharmony.

Thinking about biological space in these terms means accepting that life is neither completely portable nor fully standardizable. It is a situated event, a local equilibrium, a form that needs a "where" just as much as a "how." In an age that dreams of total delocalization—of processes, bodies, even consciousness—this awareness sounds almost like a call for caution. Or, perhaps, like a return to an ancient wisdom: one cannot separate what lives from the place that makes it possible.

Embodied biography: inhabiting a body, becoming someone.

There is a reassuring, very modern idea that personal identity resides "inside" us: in our genes, in our brains, perhaps in some deep and stable circuit. Field biology undermines this certainty. If life is information that takes shape in space, then identity, too, is not an isolated nucleus, but a situated process that stabilizes over time through environments, postures, repeated gestures, and habits. In other words: each of us is, biologically speaking, a "spatial biography."

The body is not merely what accompanies us through life; it is the place where life leaves its mark. Not in a metaphorical sense, but literally. The ways in which we sit, walk, work, and sleep; the spaces we frequent every day; the rhythms we impose on our nervous and hormonal systems: all of this contributes to organizing relatively stable fields of information that ultimately "resemble us." Identity, viewed from this perspective, is not an internal program, but a coherent configuration maintained over time.

The sociologist Pierre Bourdieu spoke of "habitus" to describe how social structures become bodily dispositions: the way one stands, occupies a room, or uses one's voice. It was a sociological insight, but today we can also interpret it biologically. Habitus is not merely learned: it is embodied. It becomes muscle tone, postural patterns, and breath regulation. In this sense, the body is a living archive that does not store data, but rather repeated spatial relationships.

Examples abound. Consider astronauts returning from *the International Space Station*. Studies conducted by NASA and the ESA have shown that, after months in microgravity, it is not just bones or muscles that change, but one's perception of oneself in space. Walking, standing still, and orienting oneself become actions that must be relearned. Not because the brain "has forgotten," but because the body's frame of reference has reconfigured itself. Motor identity—a silent yet decisive component of who we are—depends on the place where the body has resided.

This also holds true in a more everyday and less extreme sense. A contemporary dance teacher recounted in a 2016 interview how, after twenty years of practice, she recognized herself more by the quality of her movement than by her vo r thoughts. "*I know who I am by the way I walk into a room*," she said. This isn't a poetic pose: it's the description of an identity that has become established in bodily space through thousands of repeated micro-actions.

The biology of fields invites us to take these observations seriously. If the information that constitutes us depends on spatial configurations, then changing our environment, posture, or rhythm is never neutral. Moving, changing jobs, or living in cramped or overstimulating spaces does not merely mean adapting psychologically: it means slowly reshaping the informational field that holds us together. This is why certain transformations seem to "turn us into someone else," even when our beliefs remain the same.

From this perspective, personal identity is not something we possess, but something we maintain. It is a dynamic balance between body and space, between repetition and variation. And like any biological balance, it can strengthen, weaken, or lose coherence—not because of an internal flaw, but because the place where we live—physical, social, symbolic—no longer supports the form we have become.

Technology and the Redefinition of Vital Places.

For most of human history, vital spaces—those in which life organizes itself, develops, and finds its identity—were given rather than chosen. Forests, plains, cities that grew slowly around rivers and trade routes. Today, more and more often, the opposite is true: first we design the place, then we insert life into it. And this silent reversal is redefining the biological fields within which organisms—human and non-human alike—stabilize their own information.

Biotechnology, artificial environments, and digitized spaces are not merely external tools. They function as new fields of interaction, capable of influencing biological rhythms and , perception, identity, and even evolutionary trajectories. We are not facing technological "interference" in natural life, but rather the emergence of hybrid ecosystems, in which vital information is anchored to designed spaces.

A prime example is Biosphere 2, built in the Arizona desert and inaugurated in 1991. It was intended to be a self-sustaining replica of Earth, a laboratory for imagining life on other planets. The project failed in part—unstable oxygen levels, unbalanced ecosystems—but it revealed something crucial: life cannot easily be transplanted into an artificial environment without losing its coherence. Plants grew, but trees snapped because there was no wind; they had not "learned" mechanical resilience. The biological system was incomplete.

Today, artificial environments are less spectacular but infinitely more pervasive. Consider neonatal intensive care units, where premature babies spend weeks in high-tech incubators. Studies conducted in Stockholm and Boston in the 2010s showed that light, sounds, and even the regularity of medical procedures help shape neurobiological development. In these contexts, technology does not support life—it supports the structure. It becomes the primary space in which the organism organizes its own information.

Even more radical is the case of digital spaces. Here, we are no longer dealing merely with modified physical environments, but with immaterial yet biologically active places. In 2018, a group of researchers at University College London observed that prolonged use of immersive virtual environments altered bodily perception and stress response, even outside the digital experience. Once back in the physical world, the body continued to "behave" as if part of its field were elsewhere.

This is not an isolated case. A computer engineer recounted suffering from a persistent sense of disorientation after months of working exclusively in virtual reality on an industrial project. *"It was as if my body had changed its* scale- ," he said. Not a psychological disorder, but a spatial reconfiguration of bodily experience. The organism's sensory field had adapted to a space that no longer coincided with the physical body.

Biotechnology is pushing this transformation even further. Lab-grown organoids, engineered tissues, designed microbiomes: forms of life that emerge and stabilize in entirely artificial spaces. In 2013, synthetic biologist Craig Venter spoke openly about "designed life" during a conference in San Diego. He was referring not only to the genome but to the laboratory setting as the primary environment of existence. Here, the biological field does not emerge; it is deliberately constructed.

These new life-supporting environments raise unprecedented scientific and ethical questions. If biological information is inseparable from the space that sustains it, what happens when that space is temporary, proprietary, and updatable like software? Who is responsible for the integrity of an artificial life field? And above all: what kind of life can emerge in environments that have no history?

Field biology suggests a cautious answer: life can adapt to almost anything, but not without a cost. Every designed field carries with it an implicit vision of what life should be. Making this vision explicit—and taking responsibility for it—is perhaps the most urgent challenge of our technological age.

The Participating Gaze: Why Science, Too, Has a Place.

If life is information taking shape within fields of relation, then knowledge of life is also a situated phenomenon. There is no "view *from nowhere*," to use an expression dear to contemporary philosophy of science. Every act of biological observation arises within a historical, cultural, technolog , and even bodily context. Even the scientist, whether willing or not, inhabits a field.

The idea of situated knowledge did not originate in biology. It was forcefully articulated in the 1980s, particularly by Donna Haraway, who wrote in 1988: *"A vision is always a vision from somewhere*." But it is in the biology of fields that this insight takes on new significance. If the organism is not a machine that can be taken apart, but a coherent whole, then those who study it enter into a relationship with what they observe, influencing—at least in part—the way biological information manifests itself.

An old anecdote told in the hallways of the Marine Biological Laboratory in Woods Hole, Massachusetts, illustrates this point well. In the 1960s, two groups of researchers were studying the same organism—the sea urchin—to understand the early stages of embryonic development. Their instruments were similar, their techniques nearly identical. Yet their interpretations diverged profoundly. Only later was it understood that the initial questions—what they were trying to see—had shaped the entire experimental framework. It was not the embryo that had changed, but the conceptual space in which it was observed.

The biology of fields compels us to take this fact seriously: knowledge is not external to life, but a particular expression of it. Every theoretical model, every experimental protocol, every disciplinary language creates a cognitive environment that selects certain forms of information and obscures others. The laboratory, like the microscope or the algorithm, is not neutral: it is a vital place, in the fullest sense of the term.

This awareness introduces a new responsibility. If the scientific gaze is situated, then its consequences are not only epistemological but also ethical. In 1972, during the U.S. Congressional hearings on the Tuskegee study, it became clear just how much a science that believes itself to be "objective" can become blind to its own moral framework. It was n ly a matter of bad biology, but of knowledge that had lost sight of the human context in which it operated.

From a field-based perspective, the scientist's responsibility concerns not only the use of results but the very form of knowing itself. What kind of relationship do we establish with what we study? Do we treat the organism as an isolatable object or as a biological interlocutor? The difference is not rhetorical. In many contemporary ecological research practices—consider, for example, Robin Wall Kimmerer's work with Native American communities—knowledge arises from prolonged listening to a place, not from its abstraction. Here, science does not renounce rigor, but changes its stance.

A situated science, then, is not a "relativist" or weak science. On the contrary, it is a more demanding science, because it acknowledges its own limitations. It knows it operates within biological, social, and symbolic fields that it helps to shape. And it accepts that knowing life means, in part, being transformed by it.

Perhaps this is the deepest lesson of field biology: we cannot understand life without questioning the place from which we observe it. In this sense, every act of knowledge is also an act of participation. And every participation inevitably entails a responsibility.

When information becomes an organism.

There comes a moment in the study of life when the language of codes ceases to be useful. It is when we realize that form is not the final result of an informational process, but the process itself in action. Field biology proposes a decisive shift: information does not describe the organism—it constitutes it. It does not remain behind the scenes like a musical score, but manifests itself directly in the structure that holds cells, tissues, and functions together.

A classic example—one worth mentioning again, often discussed in embryology courses at Cambridge or Heidelberg—concerns Hans Driesch's experiments in the late nineteenth century. By separating the cells of an early-stage sea urchin embryo, Driesch expected to obtain "half-formed" organisms. The opposite occurred: each part gave rise to a complete, albeit smaller but coherent, individual. There was no localized plan to follow, but rather a global form that reorganized itself. The information was not located at a single point: it resided in the network of relationships that the embryo embodied.

Here, form ceases to be passive geometry. It becomes active information, capable of guiding processes, correcting deviations, and maintaining coherence. An organism does not "have" a form the way an object has a shell: it is a form that sustains itself over time through continuous exchanges of matter and energy. Aristotle had intuited this when speaking of *eidos* as an active principle; contemporary biology, using different tools, returns to that line of thought without metaphysical nostalgia.

In the 2000s, at *the Max Planck Institute* in Dresden, studies on the regeneration of flatworms revealed something surprising: decapitated animals regenerated heads with different orientations depending on the conditions of the bioelectric field. It was not the "genetic program" that changed, but the configuration of the field that shaped the future form. The head was not "written" somewhere; it emerged as the stable solution of a system.

From this perspective, form is operational memory. It "knows what to do" because it is already the result of embodied knowledge. As in an ancient city—think of Siena or Fez—the streets do not follow an abstract plan, but rather the accumulation of gestures, needs, and adaptations. Urban form contains information because it embodies it. In the same way, the organism is a story that continues to function.

Pathology also sheds light on this point. In some tumors, what is lost is not only molecular control but also belonging to a shared form. The cells still "know" how to divide, but no longer as part of a whole. The information has not disappeared: it has become decoupled from the overall form. The problem is not an excess of activity, but the loss of morphological coherence.

To say that information becomes an organism is therefore to recognize that life thinks through forms. Not in the human sense of the term, but as the capacity to select stable, functional, and meaningful configurations within a given context. Every organism is a response that has taken shape. It does not merely represent an environment: it actively inhabits it, shaping itself and shaping it.

This vision has profound consequences. If form is active information, we can no longer treat living beings as assemblages that can be repaired piece by piece. Every intervention—therapeutic, technological, or environmental—affects a formal equilibrium that carries with it implicit knowledge. Field biology does not call for resignation, but for greater attention: understanding how a form works before seeking to correct it.

After all, as Paul Valéry wrote in 1937 following a visit to a physiology laboratory in Paris, *"the living is that which cannot be explained without losing something essential.*" Perhaps that "something" is precisely this: form as a continuous gesture of information, the moment when knowledge ceases to be abstract and becomes life.

The Eye and the Field: Seeing Beauty, Sensing Life.

Anyone who has observed a diatom under a microscope, or walked along an Atlantic cliff watching the alternating patterns of brown algae, has probably experienced a familiar sensation: this form "works," and that is precisely why it appears beautiful. This is neither a sentimental judgment nor a secondary cultural reflex. From the perspective of field biology, the aesthetic experience becomes a way of recognizing biological coherence.

Natural beauty is not the same as ornamentation. It is rarely redundant. Rather, it is the perceptible sign of a successful organization, of a form that holds together multiple constraints without any obvious fractures. In 1917, D'Arcy Wentworth Thompson published the aforementioned **On Growth and* Form*, a book written between Dundee and St. Andrews, in which he demonstrated how living forms emerge from physical and mathematical equilibria even before they do from genetic instructions. Thompson did not speak of beauty, but his drawings—jellyfish, shells, bones—exerted a profound influence on artists and architects alike. A convincing form is one that needs no explanation.

Goethe, more than a century earlier, had intuited something similar while studying plants in the gardens of Padua and Palermo. His morphology did not seek abstract laws, but rather recognizable transformations: the leaf that becomes a petal, the petal that becomes fruit. The human eye—whether trained or not—perceives this continuity as harmony. Not because it is "decorative," but because it reveals a legible internal order.

Contemporary neuroscience offers an interesting piece of the puzzle. Studies conducted in London and San Diego in the early 2000s, in conjunction with the work of Semir Zeki and V.S. Ramachandran, have shown that the human brain responds positively to configurations that optimize information and simplicity without losing complexity. Living forms—a branch, a vascular network, a pulmonary alveolus—possess precisely this quality: they are complex, yet not chaotic. Our perceptual system seems to recognize, almost immediately, when a form "holds together."

From this perspective, natural aesthetics is not an evolutionary luxury, but an embodied cognitive competence. Recognizing beauty often amounts to recognizing a good integration between parts, functions, and the environment. A biologist observing a cell culture knows, even before measuring parameters, if something is wrong: he "sees" it. This visual intuition is not imprecision, but field experience. As an Italian researcher working at the Institut Pasteur in the 1990s once said, *"When a culture is healthy, you can tell by the calmness of its form."*

This sensitivity also emerges in art. Ernst Haeckel, a biologist and illustrator, published **Kunstformen der* Natur* in 1904, an atlas of marine organisms that influenced European Art Nouveau. It was not merely an aesthetic exercise: Haeckel wanted to show that living form is already an expression of order, not chaos to be tamed. Today we can interpret that work as a bridge between science and perception: beauty as the visible trace of information that has stabilized.

In field biology, all of this takes on a precise meaning. If form is active information, then our ability to perceive its harmony is a way—partial but real—of coming into contact with the functioning of living things. Beauty is not the opposite of truth, but one of its signs. It does not guarantee truth, of course, but it guides it.

Perhaps this is why living forms strike us more deeply than even the most sophisticated machines. A butterfly's wing or a fern leaf is not "perfect": it is coherent. And coherence, when it manifests itself without apparent effort, generates beauty. In that moment, the eye is not merely looking: it is knowing.

VII. When Tradition Meets Science.

There is a point in the course of this book where it becomes inevitable to confront an uncomfortable word. A word that has spanned centuries, cultures, and systems of thought, but which modern biology has progressively expelled from its official vocabulary: vital energy. This seventh chapter is situated precisely here, like a threshold. Not to resurrect ancient metaphysics nor to indulge in pre-scientific nostalgia, but to undertake a more ambitious and, in some ways, riskier endeavor: to conceptually redefine what those traditions sought to name.

Within the overall context of the book, this chapter serves as a hinge. After demonstrating, in the previous sections, the limits of biochemical reductionism and introducing the idea of life as a phenomenon of coherence and field, this chapter addresses the most delicate question: whether there is a real, observable core—translatable into scientific language—behind concepts such as *qi, r prana, and vital force*. In other words, whether what has been recounted for centuries in symbolic form is, in reality, an intuitive description of the deep organizational dynamics of living organisms.

The chapter begins with a cultural exploration. Indeed, the cross-cultural universality of the idea of vital energy is striking. From China to India, from the Greek to the Arab worlds, and on to premodern European medical traditions, the same insight consistently emerges: life is not merely matter in motion, but a dynamic order that permeates matter. The names change, the underlying cosmologies change, but the phenomenon described seems surprisingly similar. This fact, on its own, proves nothing scientifically, but it raises a legitimate question: is it plausible that entire civilizations, independent of one another, would have constructed complex concepts around a mere perceptual error?

The chapter emphasizes a crucial point: these concepts should not be taken literally, nor should they be dismissed as superstition. They should be understood as phenomenological languages—attempts to give form to real experiences—such as the vitality, resilience, and self-organizing capacity of organisms—in the absence of the conceptual and mathematical tools we have today. It is the same approach that science adopts when it reinterprets alchemy as a precursor to chemistry, or Ptolemaic astronomy as a necessary step toward a more refined understanding of the cosmos.

In this sense, the chapter engages in an ideal dialogue with a reflection by Erwin Schrödinger, who wrote as early as 1944: *"What we call life seems to behave like an ordered phenomenon kept far from* equilibrium" (*What Is Life?*). Schrödinger was not speaking of vital energy, but he clearly indicated that life cannot be reduced to a mere flow of chemical energy: it is order that is conserved, form that persists over time.

The theoretical heart of the chapter lies precisely here: translating vital energy into terms of dynamic coherence. Not a mysterious " " "force" that is added to the laws of physics, but a particular mode of organization of energy and information processes within an organism. A coherence that involves oscillations, synchronizations, and weak but structured electromagnetic fields, and that allows the living system to function as an integrated whole.

From this perspective, the words of Ilya Prigogine, Nobel laureate in chemistry, take on new relevance when he stated: *"Life is the realm of dissipative structures, where order emerges from flux.*" Here, too, there is no reference to esotericism: only the recognition that energy, in life, is never neutral but is always organized according to coherent patterns.

The chapter makes it clear that speaking of vital energy does not mean introducing unmeasurable entities. Rather, it means recognizing that biology deals with fields, relationships, and global configurations, not merely with isolated molecules. This position is echoed in the words of theoretical biologist Brian Goodwin, who wrote: *"Biological form is not imposed from the outside, but emerges from the collective properties of the system.*" If form emerges, then there is an organizing principle that precedes the individual chemical reaction.

The purpose of this chapter, within the book's overall framework, is therefore twofold. On the one hand, to defuse the false conflict between science and tradition, showing that many ancient insights can be reinterpreted without compromising scientific rigor. On the other hand, to lay the groundwork for the subsequent sections, in which health and disease will be interpreted as states of greater or lesser coherence within the biological field. Vital energy, thus redefined, is no longer a vague concept but a powerful interpretive key for understanding why an organism "functions" or breaks down.

Ultimately, this chapter does not ask the reader to believe, but to set aside prejudice. To accept the idea that science, as it advances, does not necessarily erase the past, but reinterprets it. And that perhaps, behind ancient and imprecise words, there was already hidden an acute perception of what we now call, with greater caution, biological coherence.

The word "energy": from a physical quantity to a metaphor for life.

In the 21st century, few words circulate with the same ease—and ambiguity—as the word "energy." We use it to describe a power plant, a physical formula, a person's state of mind, the vitality of an organism, or even the atmosphere of a room. It is a bridge word, spanning different languages without ever truly settling in just one. And it is precisely this ubiquity that makes it crucial to the theme of this chapter: understanding how vocabulary influences our understanding of life.

From the perspective of modern physics, "energy" is one of the best-defined concepts we have: a measurable, conserved quantity, formalized with mathematical rigor. Yet, as Richard Feynman noted in his famous **Lectures on* Physics*, *"we don't know what energy is; we only know how to calculate it."* A statement that, far from being a provocation, reveals a profound paradox: even at the heart of the most exact science, energy remains in part an operational concept, rather than an ontological one.

When this term enters the realm of biology, the ground becomes even more slippery. Twentieth-century biology inherited from physics a predominantly quantitative conception of energy: ATP, electrochemical gradients, metabolic fluxes. All of this is correct, but it is not sufficient to fully capture what we mean when we say that an organism is "full of energy" or, conversely, "lifeless." In these cases, we are not speaking merely of calories or chemical reactions, but of the quality of organization, the ability to maintain a form, and to respond coherently to the environment.

This is where traditional language—qi, prana, vital force—continues to exert a quiet pull on the collective imagination. Not because it offers better explanations, but because it names something that scientific language still struggles to make intuitive. The contemporary popularity of these terms, even in hyper-technological urban contexts like Berlin, San Francisco, or Milan, is not a return to the irrational, but a symptom of a semantic gap: there is no shared term to denote the holistic and qualitative dimension of being alive.

In the 1970s, the philosopher Ivan Illich observed how technical language, while increasing precision, can reduce lived experience to what is measurable. Something similar is happening today with the word "energy": the more we define it in a strictly physical sense, the more we risk losing the connection to the concrete experience of living beings. It is no coincidence that, in everyday language, energy is always relational: it is "transmitted," "lost," or "restored." It is a word that speaks of processes, not objects.

This ambivalence has significant cultural consequences. On the one hand, it fuels misunderstandings and oversimplifications, when energy becomes a magical explanation that fits all situations. On the other hand, however, it points to a genuine need: that of conceiving of life not merely as a collection of components, but as a dynamic, structured, and meaningful event. In this sense, vocabulary is not merely an ornament of scientific thought, but one of its most powerful tools.

The reimagining of the concept of vital energy proposed in this chapter fits precisely into this intermediate space. It does not ask us to abandon the precision of physics, but to recognize that, in biology, the word "energy" functions as an indicator of coherence—a linguistic signal pointing toward systemic phenomena, fields of relation, and collective organizations. It is a use of language closer to that of systems theory or theoretical biology than to classical physics.

As the linguist and philosopher Ludwig Wittgenstein wrote, *"The limits of my language are the limits of my world."* Applied to biology, this means that as long as we speak of life solely through the vocabulary of the machine, we will see machines. If, instead, we begin to use words that evoke processes, fields, and relationships, perhaps we will be able to conceive of the living in a way more suited to its nature.

Ultimately, the word "energy" in the 21st century is a source of cultural tension even before it is a scientific one. It is up to us to decide whether to let it slip into vagueness or to use it as an opportunity to refine our understanding of life. Redefining it does not mean emptying it of meaning, but rather restoring a new precision to it, one that matches the complexity of living things.

From popular science to pop culture: the media's portrayal of energy.

Every era tells the story of life through the means at its disposal. In our time, more than in scholarly treatises or academic circles, the idea of energy and vitality takes shape in television documentaries, science podcasts, and science festivals, but also in talk shows, digital platforms, and even the algorithms of social networks. It is here that the concept of vital energy, reformulated by contemporary science, is filtered, simplified, amplified, or sometimes distorted.

In recent decades, science communication has played an ambivalent role. On the one hand, it has helped make complex concepts—such as the organization of living systems, self-regulation, and complexity—accessible. On the other hand, it has often succumbed to the temptation of sensational storytelling, where "energy" becomes an emotional buzzword, capable of evoking mystery and power rather than understanding.

A symbolic moment marking this shift can be traced to the 1980s, when Carl Sagan, through the series **Cosmos** (1980), , brought into the homes of millions of people a vision of a dynamic universe, in which life was seen as a rare yet deeply integrated expression of cosmic processes. Sagan spoke of stars, molecules, and cells in the same narrative tone, creating an imaginative continuity between physics and biology. He did not use the language of "vital energy," but he helped construct an image of life as an organized, process-driven phenomenon that cannot be reduced to pure mechanics.

In the years that followed, however, the proliferation of media channels led to a fragmentation of the narrative. Alongside rigorous popular science—think of David Attenborough's documentaries, where the vitality of organisms emerges through behavior, ecological relationships, and cycles—a vast communications market has developed in which the word "energy" has become a narrative brand. Energy as a promise of well-being, harmony, and personal empowerment.

An anecdote shared by an Italian science journalist during the Genoa Science Festival (2016 edition) is illuminating: after a lecture on complex systems in biology, an audience member asked whether those *"interaction* networks" were scientific proof of the aura. The question did not stem from naivety, but from a hybrid imagination, shaped by simultaneous exposure to scientific and pseudoscientific content, often presented using the same visual and rhetorical language.

In fact, the media do not merely convey information: they shape cognitive expectations. When popular science uses energetic metaphors without clarifying their context, it helps maintain a gray area where different concepts—physical, biological, and experiential—overlap. The result is not necessarily misinformation, but an intuitive, emotional understanding that can be fruitful or misleading depending on how it is guided.

An interesting example comes from the world of TED Talks (a global platform where innovative and inspiring ideas are shared through short talks accessible to everyone). Since the 2000s, ideas like these have had an enormous influence on the global cultural imagination. Many talks on life, the mind, or health speak of "energy," "connection," and "vibration," often in evocative but vague terms. This language works because it responds to a real need for meaning, but it risks replacing argumentation with inspiration.

Yet, this is precisely where an opportunity arises. Mature popular science could use the narrative power of the media not to oversimplify, but to educate about complexity. Describing vital energy not as a mysterious force, but as a way to explain the dynamic organization of living organisms, requires time, patience, and trust in the public's intelligence. It is no coincidence that some of the best contemporary popular science initiatives—long-form podcasts, reportage books, in-depth documentary series—are moving in this direction.

In this sense, the role of the media is not secondary to the redefinition of the concept of vital energy: it is an integral part of the process. Because what a society believes about life also depends on how life is portrayed.

From the page to the metaverse: how the narrative of life's energy will change.

Every technological breakthrough brings with it a silent yet profound transformation: it changes the way we describe the world, and thus the way we understand it. Field biology, with its conception of organisms as stabilized informational structures, arrives at a historic moment when science communication is entering a new phase, suspended between artificial intelligence, augmented reality, and immersive environments. The question is no longer just what to say about life, but how to do so without betraying its complexity.

In recent years, artificial intelligence has become an everyday tool in the production and dissemination of knowledge. Systems capable of summarizing articles, generating scientific images, or simulating biological processes are already being used in universities, museums, and newsrooms. In 2023, for example, the *Wellcome Collection* in London experimented with interactive installations based on computational models to demonstrate the dynamics of living systems: not mere informational panels, but environments that reacted to visitors' choices. The energy of life was not merely explained, but staged as a process.

This shift is crucial. Science communication of the future will not be merely narrative or descriptive, but experiential. Augmented and virtual reality already allow us today to "enter" a biological system, to observe flows, rhythms, and synchronizations. At Stanford, a scientific visualization lab has developed VR environments for biology students in which cellular phenomena do not appear as linear chains of events, but as dynamic fields of interaction. It is a language that comes surprisingly close to the idea of vital energy as coherence, without resorting to esoteric metaphors.

Yet this communicative power carries with it a subtle risk: the seduction of excessive clarity. An algorithm tends to simplify, to smooth out what is inherently ambiguous. But life, as theoretical biologists well know, is also made up of instability, margins, and gray areas. Physicist and science communicator Philip Ball, speaking at a public event in Oxford in 2019, observed that the best scientific visualizations are not those that explain everything, but those that *"offer a glimpse of what we do not yet understand.*" This is a valuable lesson for anyone seeking to convey the energy of life through intelligent tools.

There is also another aspect, one that is more cultural than technical. Artificial intelligence is not just a tool, but a new symbolic actor. When we entrust a machine with the task of explaining the vitality of an organism, we are implicitly redefining the boundary between human knowledge and automated knowledge. An anecdote shared by a French scientific curator concerns a traveling exhibition on the biology of the future: many visitors stated that they trusted the explanations generated by AI more than those written by experts, because they were perceived as "more neutral." This is a sign that directly challenges the responsibility of science communication.

In this scenario, describing vital energy as dynamic coherence becomes a narrative challenge as well as a scientific one. New technologies can help portray life as a network, as a field, as an embodied informational process. But they require conscious cultural guidance, capable of avoiding both mechanistic reductionism and sensationalist excess.

Perhaps the future of public communication does not lie in finding a definitive definition of the energy of life, but in building contexts of understanding—spaces where the reader, viewer, or visitor can experience complexity without being overwhelmed by it. In this sense, artificial intelligence and augmented reality are not the solution, but powerful tools: like any language, they say a great deal about what we choose to tell, and even more about what we decide to leave open.

Qi, prana, life force.

There is one fact that strikes anyone who observes the history of ideas without prejudice: cultures separated by vast distances in space and time have conceived of life as permeated by an invisible dynamic principle—essential yet not reducible to matter. The names change—*qi* in classical China, *prana* in Vedic India, *pneuma* in the Greek world, *ruach* in the Jewish tradition, right up to *the "vital* force" of premodern European medicine—but the conceptual structure remains surprisingly similar. Not an object, but a process; not a substance, but a flow.

In the *Neijing*, the foundational text of Chinese medicine (dating from between the 3rd and 1st centuries BCE), qi is never defined in static terms. It is that which flows, connects, and relates organs, seasons, and emotions. Similarly, in the Indian *Upanishads*, *prana* does not coincide with physical breath, but with what gives it meaning: an organizing force that sustains life and consciousness. In Athens, Hippocrates (5th century BCE) spoke of *physis* as an internal principle of movement and regulation, an idea that echoes these Eastern visions in a surprising way.

This convergence does not stem from systematic cultural exchanges—which in many cases did not occur—but from a shared phenomenological experience: the evidence that living organisms are not random aggregates of parts, but rather dynamic wholes, recognizable by their rhythm and their ability to maintain form and function amid change. It is here that the concept of vital energy reveals its strength: not as a r causal explanation, but as a holistic description of what makes the living, living.

To dismiss qi or prana as superstitions means retroactively projecting modern scientific criteria onto eras that operated with different conceptual tools. It is a historical error even before it is an epistemological one. As the British anthropologist Joseph Needham, who devoted his life to the study of Chinese science, observed, these concepts were not naive attempts to explain the body's chemistry, but operational maps of experience, capable of guiding medical, physical, and spiritual practices for centuries.

An anecdote recounted by an Italian doctor who worked extensively in Nepal during the 1990s is illuminating: during a visit to a Himalayan village, he realized that local healers described patients' symptoms in terms of "blocked wind" or "disturbed breath." They did not speak of molecules or nerves, but they recognized with great precision states of stress, chronic fatigue, and functional disharmony. The language changed, but not the clinical focus.

The common core of these concepts is not a mysterious force that violates the laws of physics, but the intuition that life is organization in action—the maintenance of coherence over time. In contemporary terms, we might say that qi and prana are different names for what we now describe as dynamics of systemic integration: flows, synchronizations, and distributed regulatory processes that do not reside in a single point but emerge from the field of the organism as a whole.

In science, the universality of an intuition is always a sign to be taken seriously. Not proof, but a clue. The fact that civilizations with no direct contact with one another developed analogous concepts suggests that they were responding to a real structure of biological experience. As the philosopher of biology Hans Jonas wrote in 1966, life manifests itself first and foremost as "fragile self-affirmation," as a continuous effort to remain itself amid change.

Field biology fits into this long tradition, not to revive old terms, but to acknowledge that those terms pointed to something we can now reformulate with greater precision. Speaking of vital energy as dynamic coherence means translating, not contradicting. It is an act of cultural continuity, not nostalgia.

Perhaps, then, the real question is not why all cultures have spoken of a life energy, but why, for a certain period, we believed we could do without it. Or, more radically, whether the science of life is not now returning—with different tools and a different rigor—to recognize what human experience has always intuited: that life is relationship, flow, field.

The knowledge of the hands and the breath: yoga, tai chi, and martial arts as lived biology.

Long before vital energy became a theoretical issue—and even before it was criticized as a vague or pre-scientific concept—the body had already understood it. Or, rather, it was already practicing it. Yoga, tai chi, qi gong, traditional martial arts: all these disciplines share a crucial characteristic. They do not explain life; they put it into practice. In the slow movement, the regulated breath, and the seemingly still posture, what the doctrines called qi or prana takes shape as a concrete, verifiable, and repeatable experience.

It is no coincidence that many of these practices originated or took root in contexts where written transmission was secondary to oral and bodily transmission. In China, tai chi chuan developed between the 17th and 18th centuries in the villages of Henan, particularly in Chenjiagou: a family tradition rather than a philosophical one, where knowledge is passed from body to body. In India, the *asanas* of classical yoga—codified only later, betwe the 19th and 20th centuries—had existed for centuries as bodily sequences, even before they became chapters of a treatise.

Here, the language is not symbolic, but proprioceptive. As the anthropologist Marcel Mauss noted in his famous essay on the techniques of the body (1934), every culture educates its members to feel, move, and breathe in specific ways. Practices related to vital energy are, in this sense, pedagogies of perception: they train one to recognize continuity, tensions, flows, and thresholds. They do not explain what life is; they teach how it feels when it is in balance.

Watching an elderly tai chi master early in the morning in the parks of Beijing or Shanghai is an instructive experience even for a biologist. The movements are minimal, yet nothing is isolated: every rotation of the wrist involves the entire postural alignment. It is the deeply biological idea that the whole comes before the parts. The same principle runs through the yoga of B.K.S. Iyengar, who, in the 1960s in Pune, insisted on precise alignment not for aesthetic reasons, but because *"a poorly aligned body confuses the mind.*" A statement that today we might translate in terms of sensorimotor integration.

This insight is also central to Japanese martial arts. When Jigoro Kano founded judo in Tokyo in 1882, reformulating ancient techniques, he spoke of *"seiryoku* zen'yō": the optimal use of energy. This is not about muscular strength, but about coordination, timing, and economy of movement. This principle anticipates, in a practical form, what we would today call systemic optimization. Vital energy, in this context, is not a mysterious fluid, but the organism's capacity to act as a coherent unit.

These disciplines have played—often unconsciously—a role in cultural preservation. While the theoretical language of vital energy was gradually expelled from Western science between the 18th and 19th centuries, the body continued to preserve the experience of it—in monasteries, in gyms, in courtyards at dawn. A French doctor who studied qigong in the 1970s, J.-C., recounted that he came to understand the meaning of "energy" not by reading texts, but by realizing that, after months of practice, he was breathing better and getting less tired. No metaphysics: just a holistic change, difficult to pinpoint, impossible to reduce to a single cause.

This is where field biology finds an unexpected ally. These practices show that life manifests not only in measurable parameters but also in the quality of its organization. The movement becomes a window into profound dynamics: synchronization, stability, adaptation. In this sense, yoga and tai chi are not spiritual relics but somatic repositories of knowledge about life as an integrated process.

Perhaps, to truly redefine the concept of vital energy, science will have to learn not only to measure better, but also to observe differently. To accept that, at times, the body comes before theory. And that certain forms of knowledge, in order to survive, have chosen the most discreet and resilient path: that of movement.

From the temple to the supermarket: the ambiguous fate of vital energy.

There is a mirror-image risk—less discussed but equally insidious—in the attempt to scientifically redefine the concept of vital energy: its commercial trivialization. While for centuries qi and prana have been accused of being superstitions, today they face the opposite fate: they become seductive labels, printed on yoga mats, dietary supplements, and weekend "energy rebalancing" crash courses. No longer rejected, but consumed.

Walking among the booths at various wellness fairs—from Bologna to Munich, from Rimini to San Francisco—one gets the impression that vital energy has become just another commodity. Promises are made to "activate" it, "boost" it, or "unlock" it in just a few steps. This language is reassuring because it is simple, but it betrays a loss of depth. Here we are not far from what anthropologist Marc Augé would have called a cultural "non-place": concepts torn from their context, rendered interchangeable, stripped of their historical depth.

The great traditions that developed the concepts of qi and prana were by no means superficial or simplistic. In the *Neijing*, the foundational text of Chinese medicine (3rd–1st centuries BCE), the study of qi is inseparable from seasonal observation, the rhythm of the breath, and the physician's ethics. Similarly, in classical yoga texts, such as the Yoga Sutras attributed to Patañjali (2nd–4th centuries CE), prana is not an "energy to be used," but a principle to be disciplined through years of practice.

A tai chi master from Beijing, interviewed in the 1980s by a team of French researchers, responded to a pressing question with a disarming remark: *"You Westerners want to know what qi is. We learn when not to disturb it.*" It is a subtle but decisive difference. Modern commercialization, on the other hand, tends to promise immediate control, rapid results, and mastery—precisely what these traditions warned against seeking.

Distinguishing authentic cultural heritage from New Age simplifications does not mean defending an imaginary purity. Traditions live on because they transform. But certain clues are revealing. Authenticity does not promise shortcuts; it never separates work on the body from work on attention and behavior; it does not reduce vital energy to an individual resource to be maximized.

An illuminating historical example comes from early 20th-century Japan. Jigoro Kano, the founder of judo, openly opposed the transformation of martial arts into mere spectacle or a commodity. In 1922, at the Kodokan in Tokyo, he reminded his students that ki was not a magical force, but the result of coordination, posture, and intention. When a practice loses this complexity, when it becomes a slogan, we are likely no longer dealing with a living legacy, but with a caricature of it.

The paradox is clear: while contemporary biology is rediscovering, through rigorous methods, the importance of global patterns and systemic coherence, a segment of popular culture reduces those concepts to reassuring formulas. Vital energy is thus stripped of its most challenging aspect: it requires time, attentiveness, and personal transformation.

In 1998, during a conference in London on science and spirituality, physicist David Bohm—a bridge between different worlds—observed that *"the problem is not using ancient words, but using them without discipline.*" A statement that still holds true today. The scientific re-foundation of vital energy does not come through its denial or its commodification, but through patient work of cultural translation.

Reclaiming qi and prana as pre-scientific descriptions of phenomena of coherence does not mean marketing them better, but understanding them better. It means restoring their depth, freeing them from both dogmatism and marketing. In this sense, field biology offers a valuable criterion: what is vital is not what promises the most, but what maintains coherence over time.

Perhaps the true legacy of traditions is not a concept to be flaunted, but an attitude to be cultivated. In an age that measures everything in terms of performance and consumption, this is a surprisingly timely lesson.

Not magic, but experience: because vital energy has endured through the centuries.

Labeling traditional concepts of vital energy as "superstition" is often an act of cultural impatience. It means judging—using categories born in the modern laboratory—what emerged in contexts where observation was direct, embodied, and repeated over time. The civilizations that spoke of qi, prana, pneuma, or *vis vitalis* were not formulating alternative physical theories, but rather recording patterns in the experience of living beings.

Even before asking how life works, these traditions asked how it manifests itself. And what emerged, with surprising consistency, was the idea that a living organism is defined not only by what it is made of, but by how it maintains continuity, rhythm, and unity amid change. Not a metaphysical hypothesis, but a repeatedly observed empirical fact.

An often-overlooked point is that these concepts did not originate in temples, but in the trades. Qi took shape in Chinese medicine and agriculture; prana in the observation of breath and physical exertion; pneuma in Hellenistic medical practice. Galen (2nd century CE) spoke of pneuma as that which enables the body to react, adapt, and heal. It was not a mystical entity, but a descriptive function: that which is present when the body responds as a whole.

An anecdote recorded in the Royal Navy archives recounts that, during the long voyages of the 18th century, some ship's doctors distinguished sailors who were "still healthy" from those nearing scurvy not so much by obvious symptoms— —as by a sort of "general decline": posture, muscle tone, and coordination. They did not have the terminology of vitamin C, but they recognized a loss of an organizing principle. This is precisely what the concepts of vital energy sought to name.

What makes these concepts non-superstitious is their common phenomenological core. Wherever they appear, they describe three recurring aspects: the organism's continuity over time, its ability to coordinate different parts, and its sensitivity to context. They do not speak of ultimate causes, but of overall states.

Superstition, in the strict sense, attributes events to arbitrary and uncontrollable causes. Concepts of vital energy do the opposite: they restore regularity to what appears chaotic. An Ayurvedic physician from Kerala, interviewed in the 1970s by a team from the University of Madras, explained that speaking of prana served to keep the patient's overall picture in view, especially when individual symptoms changed. It was an antidote to fragmentation, not a product of it.

It is significant that these concepts emerge precisely where reductionism reveals its limitations: in healing processes, in development, and in adaptation. These are areas where the behavior of the whole cannot be immediately deduced from the parts. Field biology simply acknowledges this same difficulty, using more refined tools.

There is one final element that distinguishes qi and prana from superstition: their explanatory restraint. They do not promise miracles, nor do they guarantee total control. On the contrary, they emphasize the fragility of vital balance. In many classical texts, vital energy is not something to be accumulated, but rather something not to be squandered. This idea is surprisingly close to the modern notion of dynamic stability.

In 1932, the physiologist Walter Cannon introduced the concept of homeostasis. He did not speak of vital energy, but he described the same phenomenon: an organism's ability to remain itself through continuous fluctuations. The words change— —but the underlying experience remains. It is in this continuity that the legitimacy of traditional concepts lies.

Perhaps, then, the point is not to decide whether qi and prana "exist" in the physical sense of the term, but to recognize that they designate something that actually happens: the way in which life holds itself together. To dismiss them as superstition means losing a precious trace of our long observation of living beings.

Counting Is Not Enough: Mathematical Bias and the Shadow Cast on Vital Energy.

In modern scientific culture, there is an unwritten rule: whatever cannot be translated into numbers is, if not false, at least unreliable. It is a powerful rule that has produced extraordinary results, but over time it has turned into an epistemological bias. In this climate, concepts such as qi or prana are not even examined: they are dismissed outright. Not because they have been refuted, but because they speak a language different from that of equations.

Yet this identification of knowledge with mathematization is historically recent. Galileo, in *The Assayer* (1623), writes that the book of nature is written in mathematical language. But Galileo does not say that only what is mathematical is real; he says that it is the appropriate language for describing certain phenomena. In the centuries that followed, that caution was lost, and the method became metaphysics.

One of the side effects of this shift is the systematic confusion between the symbolic and the irrational. In everyday language, "symbolic" sounds vague, poetic—and therefore suspect. But in the history of knowledge, the symbol has always served another function: to make visible what cannot be immediately broken down.

Ernst Cassirer, a philosopher of culture active in the 1920s and 1930s, pointed out that myth, language, and science are not degrees of progressive error, but different symbolic forms. Each organizes experience according to its own criteria. To say that qi is symbolic does not mean to say that it is false; it means recognizing that it arises to guide the experience of the living, not to replace a physical theory.

An illuminating example comes from art. When Kandinsky published **The Spiritual* in Art* in 1912, he was accused of mysticism. It was only decades later that people came to understand he was attempting to describe perceptual fields, tensions, and visual resonances—not numbers, but structures. No one today would say that Kandinsky was irrational; yet he spoke a symbolic language.

In medicine, this prejudice has had concrete consequences. Michael Polanyi, a chemist and philosopher of science, spoke of *"tacit knowledge*": knowledge that guides action but cannot be fully articulated. An experienced surgeon "senses" whether tissue is healthy or not, even before being able to measure it. No one calls this superstition. Yet when similar skills are expressed through symbolic concepts, as happens in traditional medicine, the judgment changes.

In the 1950s, in Parisian hospitals, some senior clinicians distinguished between patients who were "held together" and those who were "frayed," using language that we would today call imprecise. But young doctors knew that those words often foreshadowed the clinical course. It was a holistic, unquantified description, but far from arbitrary. The problem was not a lack of rigor, but the absence of a scientific vocabulary suited to that type of phenomenon.

Perhaps the heart of modern prejudice lies here: having mistaken the tool for the ultimate criterion of reality. Mathematics is a powerful tool, but it is not neutral: it selects what is measurable and leaves in the shadows what is relational, qualitative, and dynamic. The concepts of vital energy arise precisely in that shadow, as attempts to give voice to complex configurations that defy decomposition.

As early as 1725, the Neapolitan philosopher Giambattista Vico warned against the idea that only the geometric method produces truth. *"Verum ipsum factum"*: we truly understand what we are capable of doing, experiencing, and going through. The living is experienced before it is calculated. Traditions that speak of qi or prana do not reject rationality; they reject the idea that a single form of rationality exhausts reality.

Reimagining the concept of vital energy also means this: restoring epistemological dignity to the symbolic, without turning it into dogma. The symbol is not an archaic remnant to be eliminated, but a temporary bridge between experience and theory. When field biology speaks of coherence, information, and global organization, it is performing a similar operation: it translates into a more precise language what has been expressed in symbolic form for centuries.

The real obstacle is not a lack of data, but the mental habit of recognizing as real only those things that fit into a formula. Overcoming this bias does not mean abandoning science, but expanding its vocabulary. And perhaps, as often happens in the history of ideas, the step forward consists in recovering what had been excluded too hastily.

PART FOUR. Life, Health, and Disharmony.

From the Thought of Georges Canguilhem

Health is not an ideal state to be preserved, but a creative capacity of the organism to establish new norms when the context changes.

From the thought of Henri Atlan

Living beings are not defined by the stability of their components, but by the coherence of the processes that connect them.

From the thought of Francisco Varela

Illness is not merely a local malfunction, but a rupture in the continuity through which the organism produces itself and recognizes itself as a unit.

After challenging biological reductionism and introducing the idea of life as a phenomenon of coherence and field, this Fourth Part tackles the most delicate and, at the same time, most concrete terrain: that of health and illness. It is here that theoretical hypotheses cease to be merely a new language for describing life and become a different lens through which to interpret what happens when life functions, adapts, resists, or, conversely, becomes disorganized.

In the classical paradigm, health is often understood as the absence of measurable abnormalities, while disease is equated with a lesion, a molecular defect, or a localized malfunction. The perspective of the biological fields overturns this approach without denying its usefulness: even before manifesting in tissues, genes, or clinical parameters, health and disease emerge as global states of coherence or informational misalignment within the organism.

In this framework, health is neither immobility nor perfect balance, but rather a dynamic capacity to maintain a flexible order and to absorb disturbances without losing one's functional identity. It is a continuous process of adjustment—a resilience that encompasses the entire biological field and not just its individual parts. Being healthy does not mean remaining unchanged, but rather being able to change without fragmenting.

Consequently, illness no longer appears as a specific, sudden, and isolated event, but as a progressive loss of coherence. The parts of the organism continue to exist, but they cease to interact harmoniously; the information that once coordinated the whole becomes dispersed, distorted, or localized. What emerges is a disharmony that precedes and accompanies material alterations, making them understandable as an expression of a deeper process.

This section of the book offers neither therapeutic shortcuts nor simplistic generalizations. On the contrary, it moves cautiously along a delicate boundary: on the one hand, it demonstrates the potential of a systemic and informational view of health ; on the other, it acknowledges its limitations, emphasizing the need for scientific rigor and integration with evidence-based medicine. The biological field is not a magical explanation, but an additional level of understanding.

Within the book's overall structure, Part Four represents a key transition: it translates the theoretical foundations of field biology into a reflection on concrete living, well-being, and suffering. It is the point at which life, understood as a phenomenon of coherence, reveals what happens when this coherence is maintained, disrupted, or rebuilt. And it is also the point at which biology inevitably encounters a broader responsibility: to understand health not only as an individual matter, but as an expression of broader biological and ecological relationships that make life itself possible.

VIII. Health as Coherence,

In everyday language, health is often described in negative terms: being well means having no symptoms, showing no abnormalities, and not deviating from parameters considered normal. But this seemingly reassuring definition proves fragile as soon as one observes life closely. Living organisms are never static, nor are they perfectly aligned with a fixed standard. They are constantly changing, fluctuating, compensating, and anticipating. Health, then, cannot be equated with the stability of a state, but rather with the ability to navigate change without losing coherence.

This chapter introduces a vision of health consistent with the central idea of field biology: the organism as a stabilized informational structure, capable of maintaining its identity through continuous flows of energy, matter, and information. From this perspective, being healthy means being able to integrate internal and external disturbances into a dynamic order, without the biological field losing its ability to coordinate the whole.

The concept of balance is radically rethought here. No longer a static balance, comparable to an ideal state of stillness, but a dynamic balance, similar to that of a system that regulates itself as it moves. A healthy organism is not one that does not change, but one that changes well. Informational coherence does not eliminate variation; it makes it livable.

Within the book's narrative, this chapter represents a decisive turning point: after describing life as a phenomenon of coherence and field, it demonstrates how this coherence manifests concretely in the biological experience of health. It is the point where theory meets physiology, without being reduced to it. Health thus becomes a global property of the organism, not merely the sum of properly functioning parts.

Particular attention is devoted to the resilience of the biological field: the capacity for adaptation, self-repair, and reorganization following a disturbance. Here, health appears as a distributed intelligence, not localized in a single organ or molecule, but recognizable in the way the entire system responds to challenges. It is an emergent competence, rather than a state that can be measured once and for all.

Finally, the chapter opens up implications that go beyond individual medicine. If health is field coherence, then it also depends on context: relational, environmental, and ecological. Without forcing premature practical conclusions, the need for a more systemic medicine emerges—one capable of reading the signs of health not only in isolated values but in the quality of the living organism's overall organization.

Chapter VIII translates the theoretical language of coherence and biological fields into an interpretive framework applicable to one of the most delicate concepts in biology and medicine. Lays the groundwork for the next chapter on disease as a loss of coherence, this text clarifies that health is not the zero point of pathology, but a specific mode of organization of the living.

In this sense, the chapter does not propose an alternative definition of health, but rather a shift in perspective: from the r correction of errors to an understanding of processes; from the control of parameters to the interpretation of the overall order; from the repair of parts to the care of the relationships that hold the whole together.

Health is being in sync: the invisible orchestra of biological rhythms.

If health is coherence, then its true test is time. Not the abstract time of clocks, but biological time—plural and layered—which flows through the organism like a tapestry of different rhythms. Every living being is, first and foremost, a temporal architecture: rapid and slow cycles, daily and seasonal oscillations, micro-variations that coordinate without the need for a central conductor. Health manifests itself when these rhythms—metabolic, endocrine, and neuronal—manage to synchronize, maintaining a stable relationship while remaining flexible. Chronobiology has shown that the human body functions according to an internal temporal organization that does not merely react to the environment but anticipates and interprets it. The research that led, in 2017, to the recognition of the role of the molecular mechanisms of the circadian clock has highlighted a crucial point: it is not only what happens in the body that matters, but when it happens.

A healthy organism is not one in which every process functions in isolation, but rather one in which the timing of these processes is in harmony. It is this temporal coherence that allows the whole to remain itself while changing, much like in a musical score where different themes enter and exit without disrupting the unity of the piece. From this perspective, the biological realm also appears as a field of synchronization, capable of integrating different rhythms into a dynamic order. Health, then, does not coincide with rigid regularity, but with intelligent timing: the organism's ability to be in the right moment , to attune its internal rhythms to those of the world without losing its own identity.

From the organism to the ecosystem: the same grammar of coherence.

A brief comparison between biological systems and ecosystems helps us focus, almost instinctively, on what we really mean when we speak of health as coherence. A living organism and a natural ecosystem operate on different scales, yet they seem to obey surprisingly similar organizational principles. In both cases, what matters is not the perfect efficiency of every single element, but rather the quality of the relationships that hold the whole together. A mature forest, studied by ecologists since the 1970s, is not "healthy" because every species thrives to the fullest, but because the system as a whole is able to absorb disturbances—a fire, a drought, the arrival of an invasive species—without losing its functional identity. Similarly, a healthy human organism is not one devoid of fluctuations or asymmetries, but one capable of integrating local variations into a global order.

This analogy is not merely evocative. Gregory Bateson, observing living systems and ecological systems, spoke of an "ecology of the mind," emphasizing how stability always emerges from feedback loops and not from linear control. In an ecosystem, excessive simplification—the drastic reduction of biodiversity—makes it fragile; in an organism, a similar functional rigidity compromises the capacity for adaptation. In both cases, health coincides with distributed coherence, not with the optimization of a single function.

The concept of a biological field allows us to grasp this relationship without straining the analogy. Just as a non- d ecosystem cannot be reduced to the sum of its species, an organism cannot be explained as a simple aggregate of organs or reactions. In both cases, an informational order is at work that guides local processes, maintaining a recognizable form over time. This is why, when observing certain natural environments—a lagoon, a coral reef, a stable grassland—many biologists have spoken of "healthy systems" using language surprisingly close to that of physiology. Not because an ecosystem is an organism in the strict sense, but because it shares with individual living beings the logic of dynamic coherence.

This point serves to suggest a continuity of principles: individual health and environmental health are not the same thing, but they seem to emerge from the same grammar of life, in which balance is never static and order is never imposed from above.

Healing is not fixing: the living lexicon of health.

In everyday language, "to treat" often means intervening to eliminate a problem. We treat a wound, we treat a disease, as if the body were a broken machine and health were synonymous with the removal of the defect. Yet, if we look closely at the vocabulary of care—the one that spans medicine, ecology, and even ethics—a deeper and less technical meaning emerges: to care is to take responsibility for a relationship over time. It is not a one-time action, but a continuous practice of attention, listening, and adjustment. From this perspective, field biology offers a powerful interpretive framework: health is not about eliminating unwanted elements, but about preserving the relational coherence that holds a living system together.

An illuminating parallel comes from the world of environmental conservation. When, in the 1990s, large-scale renaturalization projects began in the Rhine Delta in the Netherlands, ecologists realized that "caring for" the river no longer meant rigidly containing it with levees or forcing it into an ideal course, but rather giving it back its space, allowing its dynamics to express themselves without losing balance. Landscape maintenance thus became a process of accompaniment, not control. Something similar happens in the body: care, understood in its deepest sense, does not impose an order from the outside, but rather supports the conditions that allow the organism to reestablish its own informational coherence.

It is no coincidence that this idea resurfaces in various fields. In the 1980s, philosopher Joan Tronto defined care as an activity that encompasses *"everything we do to maintain, sustain, and repair our world.*" A world that includes bodies, environments, and relationships. Applied to health, this definition shifts the focus: it is not about fighting the body, but about collaborating with its systemic intelligence. Some primary care physicians say—often outside of official conferences—that the most effective treatment pathways begin when we stop chasing a single metric and start observing how the patient lives, works, sleeps, and breathes. This isn't romanticism: it's an implicit recognition that health emerges from a network of coherent relationships.

Traditional agricultural language also preserves traces of this vision. In many areas of the Apennines, people still speak of "caring for the land" to refer to crop rotation, the maintenance of dry stone walls, and attention to springs. You don't "repair" soil: you accompany it, season after season. Transposed to the biological realm, this metaphor becomes surprisingly precise. The organism's biological system does not require one-time, definitive interventions, but rather continuous attention and small adjustments that maintain functional relationships between its parts.

In this sense, health as coherence is also a matter of language. Speaking of care as the maintenance of relationships changes the way we think about the body and the our relationship with it. No longer a territory to be conquered or corrected, but a living environment to be inhabited mindfully. It is a quiet yet decisive paradigm shift that brings medicine closer to ecological wisdom without blurring the lines: what they share is not the object, but the organizing principle. Health, like a well-maintained ecosystem, does not impose itself. It is cultivated.

Health does not stand still: the biological art of remaining oneself while changing.

For a long time, in medicine as in common sense, health was imagined as a state of stillness: normal parameters, stable values, regular functions. A reassuring, almost geometric ideal. But contemporary biology—and even more so the perspective of field biology—suggests a less comfortable yet more fascinating truth: a healthy organism is not one that does not change, but one that changes without losing coherence. In short, health resembles a well-regulated oscillation more than a flat line.

A classic example, often cited in physiology courses but rarely discussed in depth, concerns the heartbeat. In the 1980s, some American cardiologists noted that a perfectly regular heartbeat—like a metronome—is often a sign of risk, not of health. A healthy heart exhibits continuous variability—micro-irregularities that indicate the system's ability to adapt, moment by moment, to the body's demands. Paradoxically, rhythmic immobility is a form of rigidity. In this sense, health does not coincide with the absence of fluctuations, but with their integration into a dynamic order.

This idea finds a surprising echo even outside of medicine. Evolutionary biologist Stephen Jay Gould explained, in an essay from the 1990s, how the longest-lived species are not the most specialized ones, but those capable of tolerating variations, of inhabiting a wide range of conditions without collapsing. The same holds true for the individual organism: static equilibrium—the kind that attempts to preserve an ideal state at all costs—is fragile; dynamic equilibrium, on the other hand, is a form of the system's intelligence.

From the perspective of field biology, this distinction takes on an even more radical significance. If the organism is a stabilized informational structure, its health cannot consist in the freezing of information, but in the biological field's ability to reorganize itself while maintaining a recognizable form. Like a jazz musician who improvises without losing the key, the healthy body continually varies its internal configurations without losing sight of the overall coherence. It is no coincidence that musicologist Leonard Meyer also spoke of dynamic equilibrium in relation to music: tension and resolution are not accidents, but conditions of harmony.

The history of science also offers illuminating insights. Walter Cannon, the physiologist who introduced the concept of homeostasis in the 1930s, already understood equilibrium as an active process, not a passive condition. Over time, however, his insight was simplified, transforming homeostasis into a sort of static ideal. Field biology recovers and radicalizes the original insight: balance is a verb, not a noun. It is something the organism does, not something it possesses.

In this light, health appears as a form of movement that is successfully sustained over time. Not the absence of crises, but the ability to navigate them. Not immobility, but continuity of meaning through change. A healthy organism is not one that remains identical to itself, but one that manages to remain itself while changing.

Field resilience. Adaptation and self-repair.

If biological equilibrium is not immobility, then resilience is not mere resistance. From the perspective of field biology, resilience appears rather as an organism's ability to navigate a disturbance without losing its informational form, reorganizing itself from within. It is not a passive defense, but an active, creative process that involves memory, flexibility, and a surprising economy of means.

An emblematic example—often cited in cell biology laboratories but rarely explained to the general public—concerns epithelial tissues. When a wound disrupts the continuity of a tissue, the cells do not simply proliferate randomly. Studies conducted since the 2000s, including those by Pazhuh Benoît Ladoux's group in Paris, have shown that cells "sense" the void, modify their geometry, realign mechanical stresses, and reconstruct form even before function. It is as if the organism's informational field already contained an implicit map of what needs to be restored.

This phenomenon of self-repair is not limited to the microscopic scale. After the fire that devastated part of Yellowstone National Park in 1988, many observers spoke of an irreversible catastrophe. Yet, in the years that followed, ecologists noted that the system was not "reverting" but reorganizing itself: new plant successions, different equilibria, and biodiversity that in some cases had even increased. Resilience, in this context, did not mean a return to the previous state, but rather the system's ability to maintain its identity while undergoing transformation. The same principle, on a different scale, applies to living organisms.

From a biological perspective, adaptation is not a local response to a local stimulus. It is a global response to a disturbance that is "read" by the system as a whole. Neuroscientist Kurt Goldstein, observing soldiers with brain injuries during World War I, noted that functional recovery depended not only on the damaged area but on the entire organism's ability to reorganize its behavior. It was not a matter of returning to what had been lost, but of finding a new, coherent way of being in the world. This is a lesson that resonates strongly today in field biology.

Resilience, in this sense, implies a form of memory that is not an archive but an operational trace. The body remembers not only what happened but also how it responded. Some immunologists speak of functional "imprinting": every infection, every trauma leaves a modulation in the biological field that can strengthen or weaken the body's future capacity for adaptation.

To speak of the resilience of the field means shifting the focus from damage to the living being's capacity. The organism is not fragile because it is exposed to disturbances; it is fragile only when it loses the ability to reorganize itself. Health, in this light, appears as an emergent property: not the absence of wounds, but the ability to endure them without disintegrating.

What Life Remembers: Memory, Identity, and Biological Continuity.

Every living organism is an archive. But not in the dusty sense of a static library: rather, as a living memory that selects, reworks, and utilizes the past to remain coherent in the present. In the chapter dedicated to health as coherence, the theme of biological memory becomes central because there is no continuity without memory, and there is no health without a memory capable of guiding adaptation.

The best-known form of memory is genetic memory. DNA, with its four-letter language, is often presented as the great book of life. In part, it truly is: it preserves ancient instructions, accumulated over millions of years. When Watson and Crick announced the double-helix structure in Cambridge in 1953, Francis Crick famously declared that they had "discovered the secret of life." Over time, we have come to understand that it was an important secret, but not a sufficient one. The genome is not a musical score that is always performed in the same way, but rather an open score that requires performers, context, and memory.

This is where epigenetic memory comes into play. Beginning in the 1990s, studies conducted in Sweden, in the Överkalix region, showed that food shortages or surpluses experienced by grandparents could influence the health of their grandchildren. It was not the DNA sequence that changed, but the way in which certain genes were activated or silenced. In other words, the organism "remembered" events it had not directly experienced, translating them into a regulation of gene expression. Epigenetic memory does not preserve symbolic content, but rather operational traces: instructions on how to exist in the world based on what has already happened. Yet even this memory does not fully capture the picture. There is a third dimension—more subtle and controversial—that field biology proposes to call "field memory." It is not encoded in a specific molecule, nor is it reducible to a local mechanism. It is an informational continuity that concerns the organism's overall form—the way its parts coordinate over time. A striking example comes from studies of regeneration: when a planarian is dissected, each fragment can regenerate a complete organism. But what is surprising is not just the regrowth of tissues, but the correct reconstruction of the form. How does the system "know" where the head is and where the tail is? Geneticists provide part of the answer, but not all of it. Something seems to preserve the memory of the whole. In the 1970s, biologist Rupert Sheldrake—a controversial but culturally influential figure—spoke of "morphic fields" as carriers of memory. Beyond the more speculative interpretations, the idea that form possesses its own informational inertia is now the subject of more rigorous research, for example in the studies on bioelectric gradients conducted by Michael Levin at Tufts University. Here, memory is not a repository but a condition of possibility: what allows the system to become recognizable again after a disturbance.

A general practitioner in Lyon described patients who, after technically successful surgeries, continued to "feel off" for months, as if the body needed time to update its internal image. In contrast, others, with more severe injuries, quickly regained a sense of wholeness. These clinical experiences are difficult to quantify, but they suggest that health depends not only on the repair of individual parts but also on the continuity of the body's field of meaning. Distinguishing between genetic, epigenetic, and field memory does not mean multiplying abstract entities, but rather recognizing different levels of information preservation. Genetics preserves the possible, epigenetics the probable, and field memory the recognizable. It is the latter that allows the organism to say, so to speak, "I am still myself," even after change. From this perspective, health appears as a dynamic fidelity to a form of identity that is never identical to itself.

There is a curious cultural parallel in all of this. The writer W.G. Sebald observed that memory is never a faithful reproduction of the past, but a construction that serves the present. The same holds true for living beings: they remember only what they need to continue living. Biological memory is not nostalgia; it is orientation. And it is perhaps here that the biology of fields finds one of its most fruitful points: in showing that life persists not because it resists time, but because it knows how to remember creatively.

Medical and ecological implications.

If health is coherence, as field biology suggests, then medicine can no longer limit itself to repairing isolated parts. It must examine the relationships that hold living beings together, both inside and outside the body. This shift is not a theoretical utopia: it is already underway, often in a fragmented way, sometimes unconsciously, but with concrete consequences both in clinical practice and in the relationship between humans and the environment.

In medicine, the idea of the system has quietly taken hold. A cardiologist in Stockholm recounted how some patients with heart failure improved more as a result of a change in their living environment—rhythms, relationships, sleep quality—than from yet another adjustment to their medication. This was not to deny the usefulness of medications, but to recognize that the heart does not "beat on its own": it responds to a field of nervous, hormonal, emotional, and environmental signals. Treating the organ without examining the field is equivalent, to use a metaphor dear to Anglo-Saxon clinicians, to tuning a violin while ignoring the orchestra.

In recent years, disciplines such as psychoneuroimmunology have demonstrated with solid data what clinical experience had long intuited: the immune system's state changes based on the quality of social relationships, perceived stress, and even one's sense of control over one's own life. Studies by Sheldon Cohen at Carnegie Mellon University, as early as the 1990s, demonstrated that the likelihood of becoming ill after exposure to a respiratory virus was higher in people who were socially isolated. The virus was the same, but the biological environment hosting it was not.

This perspective also has radical implications for prevention. A systemic approach to medicine asks not only “which risk factor should be eliminated?” but “which sense of coherence should be supported?” It is a shift in the question that moves the focus from waging war against the enemy to nurturing vital relationships. In some hospitals in Japan, for example, rehabilitation programs for the elderly include community activities and time spent in green spaces, based on evidence linking environmental quality to autonomic nervous system regulation. This is not Eastern folklore: it is the physiology of the field.

The same reasoning naturally extends to ecology. If the organism is a coherent system, so too is the ecosystem. Human health cannot be separated from the health of the environments in which we live. Environmental epidemiology reminds us of this with data that is increasingly difficult to ignore: the rise in respiratory diseases in the Po Valley, documented by the WHO as early as 2019, is not just a matter of individual lungs, but of chronically disharmonious environmental fields. Air, soil, noise, urban density: everything contributes to modulating biological coherence.

A telling example dates back to the water crisis in Flint, Michigan, which began in 2014. Exposure to lead did not merely affect individual organisms but produced a systemic fracture—biological, social, and cognitive. Neurological damage in children was exacerbated by stress, poverty, and distrust of institutions. It is a dramatic example of how the loss of coherence in a field—in this case, the socio-environmental field—translates into widespread pathology. No individual therapy can compensate for a degraded field.

Moving toward systemic medicine therefore also means rethinking the physician’s role. No longer merely a technician of repair, but a guardian of connections. Some family medicine programs in Canada and the Netherlands are moving in this direction, integrating clinical expertise with environmental and social knowledge. The physician returns, in an almost Hippocratic sense, to being an observer of the bios, not merely of the symptom.

Finally, there is a profound cultural implication. A medicine grounded in fields cannot promise total control or automatic healing. It does, however, offer something perhaps more realistic: the possibility of reducing the dissonances that, when they accumulate, push the system toward illness. As Ivan Illich wrote in *Medical Nemesis* (1975), the risk lies not in technology itself, but in the loss of proportion. Systemic medicine seeks to restore this proportion, recognizing that health emerges from the harmony among different levels of life.

One health, many living beings: when the biological realm transcends species.

There is an image that often recurs in the accounts of those working in the field, far from the glossy laboratories: a rural veterinarian who recognizes an impending epidemic by observing the animals' behavior before clinical symptoms appear in humans. This was already happening in the early 20th century, when the German physician Rudolf Virchow—best known for his work in cellular pathology—stated that *"there should be no dividing lines between human and veterinary medicine."* It wasn't a metaphor: it was a systemic insight ahead of its time.

Today, that insight has a specific name: *One Health.* Not a slogan, but a conceptual framework that recognizes how human health, animal health, and the health of ecosystems are different manifestations of the same broad biological field, traversed by flows of information, pathogens, chemicals, agricultural practices, and cultural choices. From this perspective, disease is no longer just an individual event, but a signal emerging from a network of interspecific relationships.

The best-known example—and one often cited only in the context of emergencies—is that of zoonoses. But reducing One Health to the issue of "species jumps" would be limiting. When the H5N1 avian influenza epidemic in 2006 alarmed the WHO and the FAO, the problem was not just the virus, but the entire structure of intensive livestock systems: animal density, genetic homogeneity, and chronic stress. The pathogen found a favorable environment, amplified by economic and environmental choices—even before biological ones.

In this sense, One Health is deeply intertwined with the biology of ecosystems: health emerges where relationships are sustainable. A conservation biologist working in the Serengeti National Park recounted how the disappearance of certain vulture species—due to the veterinary use of *diclofenac*—had led to an increase in stray dogs and, consequently, in cases of human rabies in some areas of India in the early 2000s. An event that appeared to be purely "wildlife-related" had reshaped the entire public health landscape. Once again, health did not belong to a single species.

From a medical perspective, this implies a shift in scale and responsibility. It is not a matter of asking clinicians to become ecologists, but of recognizing that every medical act has repercussions beyond the patient's body. The widespread use of antibiotics in animal husbandry—documented by the *European Medicines* Agency in its 2014 report—has contributed significantly to antibiotic resistance, now considered one of the major global threats to health. Here, the biological realm is literally shared: bacteria do not distinguish between a hospital ward and an industrial farm.

One Health also introduces a new ethical framework. In 2021, in Marseille, during an international conference on health ecology, a French epidemiologist observed that *"protecting a forest can be a more effective public health prevention measure than many drug campaigns.*" This was not a provocation, but an observation: the destruction of habitats increases forced contact between species, alters the natural reservoirs of pathogens, and destabilizes ecological systems that for centuries had functioned as biological buffers.

This vision is also reflected in the most advanced public policies. In the Nordic countries, some land-use planning programs integrate indicators of human health, biodiversity, and animal welfare into the same decision-making model. This is not green romanticism: it is biology applied on a large scale. If health is coherence, then it must be cultivated where it originates: in the relationships between living beings.

Ultimately, One Health is perhaps the most concrete expression of field biology in the contemporary world. It reminds us that we are not biological islands, but temporary nodes in a broader network. Treating humans without treating the environment and animals means intervening on a fragment while ignoring the structure that supports it. And every informational structure, when broken, sooner or later takes its toll.

The Invisible Guardians of the Field: When Health Speaks First to Shepherds.

In the lexicon of field biology—which is accustomed to discussing information, coherence, and resonance—there is a figure who rarely appears in textbooks but who persistently reappears in reality: the embodied observer. It is the herder who knows each member of his flock by name, the farmer who recognizes "tired" soil by the way it retains water, the elder of an indigenous community who senses a change in the forest before it becomes measurable. These are not poetic metaphors: they are high-resolution biological sensors, formed over the long course of a continuous relationship with the living field.

In 1972, during a severe drought in the Sahel, some anthropologists noticed that herders had begun moving their herds weeks before official meteorological indicators signaled the anomaly. They did not rely on forecasts— —but on minute details: the quality of the pasture, the behavior of insects, the way the livestock arranged themselves at night. It was as if they were reading a subtle grammar of the natural world, made up of micro-variations that escape the instrumental gaze.

This type of knowledge is not "alternative" to science: it is pre-analytical, in the noblest sense of the term. It arises from prolonged immersion in a living system and from a collective memory that integrates generations of observations. Gregory Bateson, who was certainly no nostalgic for the past, noted that the most important information in a system is often that which is not encoded in numbers. Local communities live precisely in this space: where information is still a lived reality.

An emblematic example comes from the Peruvian Andes. In the 1990s, some Quechua farmers reported an unusual increase in pests in high-altitude potato crops, which were traditionally protected by the cold. Agronomists dismissed these as isolated cases; the farmers spoke of "mountains that no longer sleep." Only years later did data confirm the rise in nighttime temperatures. Once again, the disharmony in the countryside had been sensed first in the body, in gestures, and in symbolic language.

In Europe, a lesser-known but significant episode concerns the spread of *bluetongue* in Sardinia (a viral disease of sheep transmitted by Culicoides insects) in the early 2000s. Some shepherds had noticed changes in the behavior of their sheep and the presence of insect vectors well before the epidemic was officially recognized. The initial reports were underestimated because they did not fit within standard protocols. Yet they accurately described a loss of coherence in the agro-pastoral field—not merely a veterinary problem.

What these experiences have in common is the quality of the relationship. Traditional knowledge does not separate the observer from the observed: knowledge is participation. In terms of field biology, we could say that these “ ” communities function as low-latency feedback systems, capable of detecting fluctuations before they become pathological noise. Not because they have better tools, but because they are part of the system they observe.

This raises a crucial question for health understood as coherence: ignoring this knowledge means losing part of the information available in the field. It is no coincidence that, in recent years, some environmental and agricultural health programs have begun to integrate local observations alongside satellite data. In Alaska, for example, researchers and Inuit communities have been collaborating since 2008 to monitor changes in sea ice, combining radar imagery and oral narratives. The results show that discrepancies between the two sources are often the first sign of disharmony.

Recognizing the value of this knowledge does not mean idealizing it or freezing it into romantic folklore. It means understanding that health, when it is truly systemic, requires multiple sensory channels. Field biology invites us to think of the organism—and by extension, the community—as a stabilized informational structure. Traditional knowledge is one of the most sophisticated historical forms of this stabilization: fragile, certainly, but extraordinarily sensitive.

In an era dominated by predictive models and artificial intelligence, the risk lies not so much in technology as in the loss of the ability to listen. If health is the ability to restore coherence, then the first signs of disharmony must also be sought where language is less technical but more closely attuned to life. Sometimes, the earliest diagnosis comes from those who do not speak of biological fields, but inhabit them.

Living Metropolises: When Built Space Enters the Body.

There is a moment—often imperceptible—when a city ceases to be merely a backdrop and becomes a biological agent. It happens when nighttime traffic seeps into our sleep, when the absence of trees amplifies the heat to the point of exhaustion, when constant noise reshapes the rhythms of our nervous system without asking our permission. From the perspective of field biology, the city is no longer merely a container for human lives, but a complex biological field, capable of modulating—harmoniously or dissonantly—the coherence of the organisms that inhabit it.

As early as the 1960s, urban planner Kevin Lynch observed how the form of a city influences the mental "legibility" of spaces. Today we know that this legibility is not only cognitive but also physiological. Chaotic neighborhoods, lacking visual cues and spaces for relaxation, generate a constant flood of information that the body struggles to process. It is no coincidence that, since the 2000s, epidemiological studies have shown a higher incidence of anxiety and cardiovascular disorders in high-density urban areas with poor environmental quality. In other words, the city encroaches on the individual's biological sphere.

Noise is perhaps the most obvious example of this interference. In 2011, the European Environment Agency estimated that urban noise was responsible for tens of thousands of annual cases of hypertension and sleep disorders. But what is striking, from a systemic perspective, is not only the specific harm but also the loss of coherence: the body no longer finds moments of respite; it no longer recognizes silence. The German philosopher Peter Sloterdijk has spoken of "sound atmospheres" as invisible envelopes that shape human experience. In a hyper-noisy city, that envelope becomes constantly abrasive.

At the opposite extreme, green spaces function as stabilizers. Not only because they improve air quality or mitigate heat islands, but because they introduce non-artificial biological rhythms. In 1984, researcher Roger Ulrich observed that hospital patients with a view of trees recovered more quickly than those facing a wall. Since then, subsequent studies have confirmed that the presence of urban greenery reduces cortisol levels and improves emotional regulation. It is as if the plant realm offers the human organism a memory of coherence, a reminder of more ancient patterns of life.

Some cities have begun to take this evidence seriously. In Copenhagen, as early as the 1990s, urban planning incorporated continuous green corridors, designed not as decoration but as biological infrastructure. In Milan, following the extreme summer of 2003, the issue of green spaces re-entered the public health debate even before it became a matter of aesthetics. The results have not always been consistent, but a new awareness has emerged: urban planning is a form of preventive medicine.

In this light, the city itself appears as an organism: traversed by flows, governed by rhythms, capable of either sustaining or eroding the biological well-being of its inhabitants. It is no surprise that as early as 1938, Lewis Mumford defined the city as *"a biological fact before an economic one*." Today, that insight finds new scientific grounding. Thinking about urban health, then, means asking questions not only about hospitals and clinics, but also about streets, sounds, trees, distances, and the way in which all of this shapes the body, day after day.

IX. Illness as a Loss of Coherence.

Having defined health as the ability to maintain and regenerate the coherence of one's biological field, this chapter takes a necessary and delicate step: shifting the focus to illness, not as a simple localized malfunction, but as a process of progressive disorganization. The fundamental question is not "where did something break?", but rather "when and how did the field cease to hold its parts together?"

Within the framework of field biology, illness does not immediately correspond to an observable molecular alteration. Those alterations—often decisive for diagnosis—come later. First, there is a loss of informational alignment: signals that are no longer integrated, rhythms that fail to synchronize, functions that proceed like soloists unable to listen to the orchestra. The language of pathology, as classical medicine understands it, describes the outcomes. This chapter attempts to explore the dynamic genesis of those outcomes.

This is where the notion of informational misalignment becomes central. A healthy organism tolerates fluctuations, local errors, and even temporary damage, because the overall system is able to reorganize itself. When this capacity is reduced, the parts continue to function, but no longer together. Illness thus emerges as a loss of coordination, rather than as a structural defect. From this perspective, pathology is not a single event, but a trajectory: something that unfolds over time, often long before it is named.

This is why the chapter emphasizes the idea of disease as a process—a process that can accelerate or slow down, stabilize or degenerate, depending on the field's residual resilience. This view does not deny the importance of clinical observation or biochemical explanations; rather, it places them within a broader framework, in which molecular data becomes a belated snapshot of a story that began elsewhere. It is a perspective that requires conceptual rigor, because the risk of oversimplification is real: speaking of a "field" does not justify therapeutic shortcuts or vague explanations.

The purpose of this chapter, within the broader context of the book, is precisely this: to demonstrate the logical continuity between health and disease, avoiding the treatment of them as opposing and separate states. If health is dynamic coherence, illness is a coherence that narrows, stiffens, or fragments. Understanding this transition is essential to laying the groundwork for the subsequent chapters, in which life will be interpreted as a cognitive process and as an event of meaning, even in its moments of crisis.

Certain voices from the history of medical and scientific thought help guide this reading, without the need for forced interpretations. Georges Canguilhem, a philosopher and physician, wrote in 1943: *"Illness is not a quantitative variation of the normal, but a new norm of life."* A phrase that resonates powerfully with the field's perspective: pathology is not simply subtraction, but reorganization—often a poorer form, yet still organized.

René Leriche, a 20th-century surgeon and medical thinker, offered an equally illuminating definition: *"Health is life in the silence of the organs*." When that silence is broken, there isn't always immediate pain or injury: sometimes there is only background noise, interference, or a loss of harmony.

Finally, drawing on the physics of complex systems, Ilya Prigogine noted: *"Life is possible only far from equilibrium."* If life exists under conditions of controlled instability, then disease can be understood as the failure of that control, not as a simple return to a simpler order.

This chapter thus serves as a conceptual threshold: it invites the reader to abandon the idea of disease as an external enemy or a sudden accident, and to consider it as a language of the field—sometimes obscure, sometimes open to misinterpretation, but rarely random. Without romanticism or easy promises, the field-based perspective offers a shift in viewpoint: from repairing the part to understanding the dynamics.

When Cells No Longer Understand Each Other: The Silence That Precedes Disease.

At the heart of Chapter IX, the theme of informational misalignment perhaps represents the most radical—and at the same time most concrete—aspect of field biology. Here, disease is observed at its most subtle and least spectacular moment: when the parts of the organism continue to function but cease to truly communicate with one another. There is not yet a "diseased" organ, nor is there an off-scale parameter. Rather, there is a growing difficulty in integration, a loss of a common language.

From this perspective, the organism has never been a sum of isolated units, but rather a community of communicating processes. Hormones, electrical signals, metabolic fluctuations, circadian rhythms: everything contributes to maintaining a sort of ongoing conversation. Information misalignment emerges when this conversation becomes fragmented. Messages arrive late, or with an intensity that is no longer appropriate; responses are correct locally, but inappropriate for the overall context.

An example familiar to experts but rarely discussed outside specialized circles concerns the studies on biological rhythms conducted since the 1970s. When Franz Halberg, a pioneer of chronobiology, observed that minimal alterations in the synchronization of rhythms could precede metabolic or immune disorders, he was not speaking of a direct cause, but of a signal of internal discord. The system wasn't broken—it was out of sync.

This "being out of sync" is a good metaphor for informational misalignment. The parts are not damaged, but they no longer share the same measure of time, space, or function. In field biology, this is equivalent to saying that the informational field can no longer maintain effective coordination. At this stage, the disease is not visible as an object but is perceptible as a loss of orientation.

This is not an entirely new idea. As early as the mid-19th century, Claude Bernard had already intuited that the stability of the organism depended on the quality of its internal relationships rather than on the robustness of its individual elements. In a lecture in 1865, he noted: "*The stability of the internal environment is the condition for free life.*" When this dynamic stability is disrupted, the organism's freedom is reduced, and with it its ability to adapt. Informational misalignment arises precisely when differences no longer produce appropriate effects, when signals do not generate coherent responses.

Neurophysiology also offers eloquent examples. In the 1990s, studies on epilepsy showed that seizures do not simply result from neuronal hyperactivity, but from a loss of coordination between networks. Neurons fire, but not together; they communicate, but without a shared context. Once again, the problem is not energy, but its informational organization.

The value of this perspective, in the context of the book, lies in defusing the idea of illness as a sudden betrayal by the body. Informational misalignment suggests a slower and more comprehensible story: a series of micro-misunderstandings that, when accumulated, make the organism increasingly less capable of responding as a whole. Pathology thus becomes a consequence, not a starting point.

It is important to reiterate that this is not a consolatory interpretation. Recognizing the misalignment does not mean we can always correct it. But it does mean restoring an intelligible dimension to illness, freeing it from both biological fatalism and therapeutic oversimplifications. The system is not "cured" by a single, act, but is reoriented—when possible—through processes that require time, context, and moderation. As the physiologist Walter Cannon observed when introducing the concept of homeostasis in 1932: *"The body possesses a wisdom that operates beneath consciousness*." Informational misalignment is perhaps one of those moments when that wisdom struggles to make itself heard.

Illness does not simply happen: it takes shape.

There is a deep-rooted habit, in medicine as in common sense, of thinking of illness as an event—a before and an after. A test that "turns positive," a symptom that "appears," a diagnosis that "falls" upon the patient's life like a date on the calendar. But field biology—and, more generally, a dynamic view of the organism—calls for a radical shift in perspective: pathology does not simply happen; it develops.

Viewing disease as a process means shifting the focus from the moment something breaks down to the long period during which something transforms. The loss of coherence is not a sudden accident, but a gradual drift. A slow, often imperceptible slip out of balance, in which the organism continues to function while the way it functions changes.

A prime example comes from oncology. In the 1990s, Bert Vogelstein and his team at Johns Hopkins University demonstrated that colon cancer was not the result of a single mutation, but of a sequence of progressive alterations spread out over time. Not a stroke of bad luck, but a biological history. Each step was necessary but not sufficient; each phase opened up possibilities and closed others. In this interpretation, the tumor is a process of pathological reorganization, not an object that appeared suddenly.

This idea has illustrious precedents. In 1913, William Osler—one of the fathers of modern clinical medicine—reminded his students in Baltimore that *"medicine is a science of uncertainty and an art of probability."* These words still re ly today as an invitation to recognize the intrinsic temporality of disease. If pathology is a dynamic process, then a diagnosis is always a partial snapshot, taken in the midst of movement.

Field biology reinforces this insight: what evolves is not just a lesion or a parameter, but the organism's entire informational structure. Chronic inflammation, for example, is not simply inflammation that lasts too long, but a system that has learned—so to speak—a rigid, repetitive response, incapable of returning to its baseline state. The problem is not the intensity of the signal, but its persistence over time.

An anecdote often recounted in gastroenterology departments concerns the discovery of *Helicobacter pylori*. When, in 1982, Barry Marshall and Robin Warren hypothesized that many ulcers were the result of a chronic infection and not merely stress or hyperacidity, they were met with skepticism. An ulcer was viewed as an acute event. In reality, it was a slow process of interaction between the bacterium, the mucosa, and the immune response. The pathology was not the hole in the stomach, but the history that had made it possible.

Conceiving of disease as a process has profound consequences on a conceptual level as well. It means recognizing that there is no absolute starting point for pathology, just as there is no static equilibrium of health. There are trajectories. Curves. Trends. When the biological system loses coherence, it does not collapse: it reorganizes itself into forms that are less functional, sometimes more stable, but biologically poorer.

In 1977, during a conference in Brussels, the theoretical biologist Conrad Waddington used an effective metaphor: *"Development and disease are two different ways of rolling down the same hill."* In this image, pathology is not a fall, but a progressive deviation from the path.

Limitations and Possibilities of the Holistic View.

Every paradigm shift brings with it a temptation: that of explaining everything—and too quickly. The field theory approach applied to biology—and in particular to disease—is no exception. Precisely because it is powerful, evocative, and capable of unifying disparate phenomena, it requires a more rigorous conceptual discipline than usual. Recognizing disease as a loss of coherence within the biological field does not justify therapeutic shortcuts or oversimplifications.

The limits of the field theory are not a flaw, but a sign of maturity. The first limit is epistemological: a biological field is not an object that can be isolated, weighed, or photographed like a tissue or a molecule. It is a description of order, a way of interpreting the relationship between processes. To expect this theory to offer the same linear predictability as biochemistry would be a category error.

Karl Popper, in a 1965 lecture in London, noted that *"science does not progress by accumulating certainties, but by eliminating errors."* Applied to field biology, this means distinguishing between what the model can clarify—global dynamics, loss of integration, thresholds of instability—and what it cannot yet guarantee: direct therapeutic protocols, recovery times, and individual outcomes.

The historical risk is well known. Whenever concepts such as "field," "energy," or "information" leave the laboratory without adequate caution, they end up being co-opted by the language of easy wellness. This happened repeatedly in the twentieth century. In the 1930s, in Vienna, theories about morphogenetic fields were discussed in serious scientific circles, but almost simultaneously, popular versions were circulating that promised vague and universal cures. The problem was not the idea itself, but its careless use.

The potential of a field-based perspective, however, lies elsewhere. It lies in the ability to offer an interpretive framework that holds together diverse data without reducing them to a single cause. Georges Canguilhem, a philosopher and physician, wrote in 1966: *"Illness is not a quantitative variation of health, but a new norm of life*." Field biology takes up this insight and reinterprets it from a systemic perspective: when the field reorganizes itself in a dysfunctional way, the organism is not "less alive," but alive according to a poorer form of coherence.

Rigor, in this context, consists of keeping the interpretive, descriptive, and therapeutic levels separate. Field biology can help us understand how a pathology emerges and becomes established; it cannot, on its own, dictate how to intervene in every single case. Here, dialogue with evidence-based medicine is not optional, but necessary.

Thomas Kuhn observed in 1962 that *"a new paradigm does not immediately solve all problems: it changes the kind of problems worth asking*." The most fruitful potential of the field-based perspective lies precisely in this: reformulating the questions. No longer "where is the malfunction?", but "what pattern has emerged?". Not "how can the symptom be eliminated?", but "what dynamic keeps it stable?".

Accepting the limits of the field-based perspective means protecting it from ideological drifts and, paradoxically, making it more incisive. It is in the rejection of easy promises that this perspective demonstrates its scientific rigor. It offers no shortcuts, but horizons of understanding.

Thinking in terms of fields, without omnipotence: what biology can learn from the planet.

Every time a science stops looking at isolated objects and begins to think in terms of fields, systems, and relationships, it gains depth but loses an illusion: that of total control. It is a lesson that field biology is still learning, but one that ecology and climatology have known well for decades. And it is precisely from their rigor—often unspectacular but fruitful—that the field-based approach in medicine can draw a reliable compass.

Modern ecology officially emerged after World War II, when scholars such as Eugene Odum, at universities in the southern United States, began describing ecosystems not as collections of species, but as networks of energy and information flows. Odum emphasized a point that is surprisingly relevant today: an ecosystem cannot be predicted in detail, but its trends can be understood. Those who asked him for precise models, specific answers, or quick solutions were disappointed. But those who accepted the uncertainty discovered something more solid: the limits beyond which a system collapses.

A famous anecdote recounts that, during a conference at Woods Hole in the 1960s, Odum replied to a young researcher who asked how to "fix" a degraded wetland: *"First ask yourself whether you are part of the problem or part of the system."* It was not a philosophical quip, but a methodological reminder. Field ecology works because it doesn't promise simple solutions, but calls for slow, multilevel, and often uncomfortable diagnoses.

Climatology has followed a similar path. When the first large-scale global climate models were developed in the 1980s, none of the scientists involved—from Syukuro Manabe to James Hansen—thought they could "cure" the climate. The language was cautious, almost humble: scenarios, probabilities, confidence intervals. The strength of the IPCC, founded in

1988, lies not in offering immediate solutions, but in having built a rigorous consensus on the limits of the Earth system.

Here, the parallel with field biology is illuminating. Even in the human body, as in the climate, field-based thinking is fruitful only if it accepts that it cannot be immediately translated into practical solutions. Climatology has resisted—not without difficulty—the temptation to transform itself into a salvific technology. It has not promised to "cool the planet," but rather to understand critical thresholds, points of no return, and slow yet irreversible dynamics. And it is precisely this rigor that has made it scientifically credible.

Another example comes from the concept of resilience, introduced into ecology by C.S. Holling in 1973. Holling showed that an ecosystem can appear stable even as it dangerously approaches a threshold of collapse. Stability does not mean health; just as in biology, a pathological state can be highly resilient. Field ecology has never confused resilience with invulnerability. On the contrary, it has taught us to be wary of systems that seem to function "too well."

The potential of the field-based perspective in biology lies here: in changing the nature of the questions, not in providing immediate answers. Ecology and climatology have learned that thinking in terms of fields means, above all, learning not to intervene in the wrong way. It means recognizing when a local action produces unpredictable global effects. It means distinguishing between understanding and control.

There is a symbolic location that embodies this scientific maturity: Antarctica. There, at research stations like McMurdo or Concordia, scientists study extreme systems, knowing that every piece of data is partial and every model is incomplete. No one talks about "curing" the ice. The focus is on monitoring, interpreting, and anticipating. It is a science that advances by dispelling illusions, not by piling up promises.

Field biology, if it wishes to avoid easy therapeutic missteps, must take the same step of humility. It must accept that a biological field is not a switch to be flipped on, but a dynamic equilibrium to be understood. The disciplines that have made systems thinking their cornerstone teach a simple yet stern lesson: rigor arises from respect for limits.

PART FIVE. Beyond Reductionism.

From the thought of Gregory Bateson:
"The mind is not confined to the brain, but is immanent in the circuits that connect the organism and the environment."

From the thought of Francisco Varela:
"To live is to know, and to know is a way of being."

From the thought of Hans Jonas:
"In the phenomenon of life, being exposes itself for the first time to the risk of meaning."

From the thoughts of Erwin Schrödinger:
"Life seems to elude known laws, but perhaps this is a sign that our laws are incomplete."

After navigating the crisis of the classical biological paradigm, exploring life as a phenomenon of coherence, and understanding health and disease as states of the biological field, the book's journey reaches an inevitable point: rethinking what we mean by life, knowledge, and scientific responsibility. Part Five does not simply add new concepts but shifts the vantage point. It is here that field biology reveals its deepest implications.

"Beyond reductionism" does not mean rejecting analytical science, nor does it mean devaluing the contribution of molecular biology or the neurosciences. It means recognizing that, once we accept that organisms are stabilized informational structures, we can no longer consider cognition as an evolutionary accident that appeared late in the evolution of the human brain. Rather, cognition becomes a pervasive property of living beings—a way in which life orients itself, assesses its environment, anticipates, and maintains its own coherence in the world.

Chapter X stands precisely at this threshold. Here, knowledge does not coincide with reflective thought, nor with language or consciousness as we commonly understand them. To know, from this perspective, means being sensitive to the differences that make a difference, responding appropriately, and building an active relationship with the environment. Every organism, from the simplest to the most complex, knows because it lives, and lives because it knows. It is a silent yet radical reversal that removes cognition from the exclusive domain of humans without, for that reason, trivializing it.

Chapter XI takes this movement a step further, addressing the cultural and epistemological consequences of field biology. If life is not merely organized matter but a meaningful process, then science, too, is called upon to redefine its own status. Models change, of course, but above all, the legitimate questions change: no longer just "how does it work?", but rather "what kind of being is what we study?". In this space, a renewed dialogue opens up between science and metaphysics—not as a return to mystical explanations, but as a recognition that all knowledge of the living implies an ontology, whether explicit or implicit.

This part of the book thus serves as the final hinge of the entire journey. Having shown that life cannot be reduced to reactions, that the organism precedes its parts, and that health and disease are states of field coherence, it becomes inevitable to question humanity's place in this scenario. No longer an external observer, nor a dominator of living beings, but a participant in broader cognitive processes, responsible for the images of life it produces and disseminates.

The purpose of this Part Five is not to offer definitive answers, but to establish a new ground for discussion—a ground where biology can remain rigorous without being reductionist, open to meaning without losing precision. It is here that field biology ceases to be merely a theory of the living and inevitably becomes a reflection on humanity, on human knowledge, and on the way humans inhabit the world.

X. Life as a Cognitive Process.

This chapter marks a decisive conceptual threshold in the book's journey. After demonstrating that life is not limited to chemistry, that the organism is a coherent whole, and that biological fields organize embodied information, an inevitable consequence emerges here: to live is already to know. Not in the lofty, reflective sense we attribute to human thought, but in a more primordial, more sober, and at the same time more radical sense.

The thesis running through this chapter is simple only in appearance: cognition does not arise with the brain, but accompanies life from its most elementary forms. Every organism, by the very fact of maintaining its coherence over time, discriminates, selects, responds, and anticipates. In other words, it measures the world and itself within the world. This activity is not a functional addition, but the very heart of the life process.

While in previous chapters life was described as a stabilized informational structure, here information reveals its dynamic nature: not merely form, but action. Not merely order, but orientation. Field biology cannot stop at the description of configurations; it must ask how these configurations are able to persist, adapt, and transform.

When we speak of "knowing without thinking," the reference is not to a poetic metaphor, but to a biological observation: an organism's sensitivity to the conditions of its environment is not passive. It is an active, structured, selective relationship. The response is never purely reactive; it always contains an element of anticipation. In this sense, the organism appears as a system that measures—not because it possesses instruments, but because it is itself a living measuring instrument, continuously calibrated for its own survival and coherence.

This idea shifts the focus of biology. Knowledge is no longer an event that occurs within a separate subject, but a process that emerges between the organism and its environment. Cognition becomes a relational, distributed, embodied property. It is here that the chapter opens, cautiously yet decisively, toward a possible biology of consciousness: not as a reductive explanation of experience, but as an expansion of the conceptual framework within which experience itself becomes conceivable.

This Chapter X sets the stage for the work's final section. By showing that cognition is an intrinsic dimension of life, it makes the question that follows inevitable: what kind of science is possible when the object of study is also, to some extent, a subject? Field biology does not lead to a dissolution of disciplinary boundaries, but to their responsible renegotiation. Here the discussion of life as a natural phenomenon comes to a close, and the discussion of life as a bearer of meaning begins.

When information is life, not computation.

At this point, the distinction between computable information and lived information becomes decisive for understanding what makes a living organism radically different from any artificial system, even the most sophisticated one. Computable information is that which can be formalized, encoded, and transferred without loss of meaning: it is the information of codes, algorithms, and sequences that can be processed indiscriminately by a machine today or tomorrow, here or elsewhere. It is the same information that, in 1948, Claude Shannon explicitly defined as independent of meaning, focused on the efficient transmission of signals.

Lived information, on the other hand, never exists "in the abstract": it is always situated, embodied, and imbued with value for those who receive it. An organism does not process data, but rather differences that matter for its own survival and coherence. Gregory Bateson spoke of "a difference that makes a difference," and it is no coincidence that this phrase has returned today to the center of the debate on biological cognition. A thermometer records a temperature variation; an organism feels heat or cold as something that affects its internal balance.

Even the most recent artificial intelligence systems, capable of learning and adapting, remain confined to a realm of information without existential stakes: they can optimize a function, but they risk nothing. As Hans Jonas observed as early as the 1970s, *"in metabolism, form is maintained only by putting its own matter on the line*": it is this continuous risk that transforms information into experience.

It is no coincidence that when Norbert Wiener, the father of cybernetics, reflected on the limits of machines, he admitted that they "do not pay the price for their decisions." The organism does. And it is precisely this price—vulnerability, finitude, exposure to the world—that makes biological information something that cannot be reduced to calculation, because it is always intertwined with a point of view, a history, a body.

To live is to be ahead of the world.

To speak of cognition without addressing time would mean leaving the picture incomplete. For life never perceives in the pure present: it always perceives between what has already happened and what might happen. Anticipation and memory are not ancillary functions, but the two poles through which the biological realm maintains its coherence. An organism lives because it manages to remain in time without dissolving, and to do so it must continually anticipate, remember, and adapt.

Biological cognition is not limited to reacting to stimuli: it is structurally projected forward. Even the simplest cell does not wait for the environment to overwhelm it; it prepares itself, orients itself, and modifies its response thresholds. In this sense, life is always a little ahead of itself. The American neuroscientist Dean Buonomano has written that *"the brain is not a machine for thinking, but a machine for time*": a statement that can be extended to the entire organism. Even before neurons, it is metabolism that is temporal, made up of cycles, rhythms, and windows of opportunity. Anticipation is not a conscious prediction, but a dynamic structure of the field—a tension toward the possible that allows living beings to never be completely at the mercy of chance.

On the other hand, biological memory is not the same as an archive of memories. It is a diffuse memory, inscribed in the forms, thresholds, and operational preferences of an organism. Every experience leaves a trace that alters the field, making it slightly different from before. Here, time is not reversible: what has been experienced changes the way a living being will be able to respond in the future. Biologist and Nobel laureate Gerald Edelman spoke of "Darwinian memory" to describe this process of continuous selection of effective responses. It is not a matter of preserving the past, but of incorporating it. The coherence of the field arises precisely from this continuous integration between what has been and what is expected.

This view finds a surprising echo in Henri Bergson's philosophy of time, which, as early as the beginning of the 20th century, distinguished the time of physics from the time of life. Bergsonian durée is not a succession of instants, but a qualitative flow in which past and present interpenetrate. An organism does not pass through time like a moving object: it is time that organizes itself. Even contemporary biology, when studying circadian rhythms or physiological plasticity, seems to approach this insight: time is not merely an external framework, but an internal dimension of the living.

From this perspective, the biological realm can be understood as a stabilized temporal structure, capable of sustaining itself only because it manages to hold memory and anticipation together. When this integration breaks down, coherence is lost. It is no coincidence that many forms of biological and psychological suffering involve a disruption of lived time: an inability to learn from experience or, conversely, excessive rigidity that prevents adaptation.

The brain is not enough: where does cognition truly reside?

In the course of the book, this chapter marks a delicate and, in some ways, provocative turning point. After demonstrating that life cannot be reduced to a biochemical machine or to the sum of its parts, the chapter now touches on one of the raw nerves of contemporary science: the idea that cognition is identical to what happens in the brain. Neuroscience in recent decades has produced extraordinary results, mapping circuits, functional areas, and the neural correlates of experience. But this very success risks becoming a limitation when the brain is treated as the sole legitimate locus of cognition.

This critical examination is not an indictment, but rather a shift in perspective. Locating cognition exclusively in the neurons implies assuming that it is an emergent function of a specialized organ. But what happens if we take seriously the hypothesis that every organism is capable of cognition, even without a nervous system? In the 1990s, in the plant biology laboratories at the University of Bonn, researchers observed that plant roots were able to distinguish between sounds and vibrations in the soil, directing their growth toward distant water sources. No brain, no synapses, and yet a form of decision-making. As the Chilean neurobiologist Francisco Varela provocatively noted, *"Cognition is not what the brain does, but what an organism does in maintaining relationships*."

Classical neuroscience, born of a long tradition of localization dating back to Paul Broca and his Parisian studies on language loss (1861), has constructed a geography of the mind—useful, powerful, but partial. The field of biology, on the other hand, suggests a different picture: cognition as a distributed property that emerges from the coherence of the entire organism within its environment. It is not the brain that "contains" cognition, but rather the brain that participates in a broader cognitive process involving posture, metabolism, immune systems, and internal rhythms. Neuroscientist Antonio Damasio acknowledged this with intellectual honesty when he wrote that *"the brain does not think on its own*": it needs a body that feels and a world with which to interact.

When cognition ceases to be confined within the skull, our understanding of experience also changes. It is no longer a matter of representing the world, but of actively engaging with it, of resonating with it. In this sense, the biological field functions as an integrative system that precedes the distinction between sensation and response.

This shift has profound consequences. If cognition is a field process, then consciousness, too, cannot be reduced to a neural epiphenomenon.

From the Heart to the Brain: A Western History of Localized Thought.

The question is not only what cognition is, but also why we have come to look for it precisely where we now believe we can find it: in the brain. Its rise to the status of the "sovereign seat" of the mind is neither a natural nor an inevitable development. It is a cultural history, shaped by theoretical choices, clinical successes, technological metaphors, and, not least, powerful symbolic images that have guided our way of thinking about life for centuries.

In antiquity, the picture was different. For Aristotle, who taught at the Lyceum in Athens, the center of psychic life was the heart: warm, pulsating, responsive. The brain played a secondary, almost cooling role. This view was not naive, but consistent with a direct experience of the body, in which emotion, decision-making, and movement appeared inseparable. Even in the Roman world, Galen, a physician to the gladiators in Pergamon (2nd century A.D.), while recognizing the importance of the brain through his study of injuries, maintained a distributed view of vital functions.

The turning point came with modernity, and it has a symbolic date: 1641, when René Descartes published the **Meditationes de prima philosophia**. Here, the mind becomes **res cogitans**, separate from the body, **res extensa**. The problem is not only dualism, but its practical consequence: if the mind is something distinct, then a seat must be found for it. Descartes locates it in the pineal gland, a precise point, almost a control center. It is a foundational move: cognition becomes localizable. From that moment on, the mind can be sought out, dissected, and studied as an object.

Nineteenth-century medicine consolidated this trajectory. In 1861, in Paris, Paul Broca presented to the Société d'Anthropologie the famous case of "Tan," a patient unable to speak but capable of understanding. Upon his death, a e lesion in the left hemisphere appeared to explain the disorder. The idea that complex functions could be mapped to specific areas of the brain was irresistible. It was effective, produced clinical results, and reinforced the authority of the nascent field of neurology. As John Hughlings Jackson would later write in London, "localization is a method, not a metaphysics," but this distinction would often be forgotten.

The twentieth century added another decisive piece to the puzzle: technology. With the advent of the electroencephalogram in the 1930s, and then neuroimaging techniques in the late 1970s, the brain became visible. The colorful images from functional magnetic resonance imaging, displayed at conferences and in newspapers, exerted an enormous fascination. What one sees seems real—more real than what one hears or experiences. At a famous conference in San Diego in the early 2000s, a young researcher presented brain maps of attention as if they were photographs of the experience itself. Someone in the audience murmured, *"Finally, we can see thought.*" It was a joke—but not entirely.

At the same time, computer science provides the defining metaphor. Since the middle of the last century, the brain has been described as a biological computer: input, processing, output. In this framework, cognition can only reside in the "central processor." It is a powerful metaphor, but a reductive one. As the philosopher Gilbert Ryle observed with irony as early as 1949, speaking of the mind as a separate entity is a *"category* mistake." Yet the metaphor works because it is compatible with a technological culture that prioritizes control, prediction, and decomposition.

This is how the brain becomes sovereign: not because it explains everything, but because it lends itself to being better explained with the available tools. The price of this success is a gradual marginalization of the organism as a cognitive whole. Chapter X serves as a critical intervention here: not to deny the brain's role, but to restore its place within a broader field. If life is a cognitive- l process, then cognition precedes its neuronal specialization. The brain is not the origin of knowing, but one of its forms that has emerged historically and biologically.

In this sense, field biology does not propose an iconoclastic revolution, but rather a correction of perspective. It invites us to recognize that the idea of a mind confined within the skull is the result of a specific, situated, culturally oriented history. And that perhaps, today, we are ready to tell a different story.

We also think with our gut: cognition beyond the brain.

If we have come to recognize life as a cognitive process, delving into peripheral systems is the point at which this assertion ceases to be abstract and becomes almost inevitable. Because we need only shift our gaze just outside the brain to realize that cognition is already there—distributed, active, and often faster and more reliable than conscious cognition. The organism knows before it thinks, and it does so with its entire body.

Take the gut, for example. For centuries relegated to the role of a mere digestive tract, today it is at the center of a veritable scientific revolution. In the late 1990s, in the laboratories at Columbia University, neurogastroenterologist Michael Gershon began speaking openly of a "second brain." This was no journalistic metaphor: the enteric nervous system contains hundreds of millions of neurons, produces neurotransmitters such as serotonin, and makes autonomous decisions. But the real novelty isn't quantitative. It's qualitative: the gut evaluates, discriminates, and learns, in constant dialogue with the microbiota and the environment. It does not think, but it knows. And above all, it guides the behavior of the entire organism, influencing emotions, attention, and even complex choices.

From the thoughts of Antonio Damasio: *"The body is not the brain's support; it is its constant interlocutor.*" This sentence encapsulates a conceptual shift: cognition is not a central, " " control tower, but a network of systems that gauge the world from different perspectives. The gut assesses the quality of the chemical environment; the endocrine system translates overall states into slow but powerful signals; the immune system distinguishes between what is familiar and what is foreign. All these processes have a cognitive structure, because they involve recognition, memory, and selective response.

The immune system, in particular, is perhaps the most surprising example. In the early 2000s, studies conducted by the Pasteur Institute in Paris and Stanford University showed that immune cells do not simply react to an "enemy," but over time construct a veritable history of interactions. They remember, anticipate, and modulate the intensity of the response. Immunologist Polly Matzinger, with her *"danger theory,"* has proposed that the immune system responds not so much to *the non-self* as to contextual danger signals. (In immunology, "non-self" refers to anything that does not belong to the organism, such as pathogens or foreign molecules.) In *the danger theory,* the immune response is guided more by signals of damage than by the simple distinction between "self" and "non-self."

The hormonal system also participates in this diffuse intelligence. Hormones do not transmit specific information, but rather configurations of meaning: states of alertness, readiness, withdrawal, and openness. A Milanese endocrinologist described how some patients, even before developing obvious clinical symptoms, exhibited *"changes in bodily atmosphere"*: unexplained fatigue, sleep disturbances, subtle shifts in emotional tone. These signals could not be pinpointed, but they were coherent. The body was already speaking, long before the diagnosis was made.

This perspective radically changes our relationship with neuroscience. It does not refute neuroscience, but it does shift its focus. The brain remains a crucial node, certainly, but it is not the absolute ruler. Rather, it is an organizer of coherence, an advanced interpreter of signals that originate elsewhere. When cognition is viewed as a property of the biological field, peripheral systems are no longer " " "accessories," but rather a constitutive part of knowing. As the French physiologist Claude Bernard wrote as early as 1865, *"the stability of the organism is an active achievement.*" Today we can add: it is a cognitive achievement.

In this sense, speaking of organisms as stabilized informational structures means recognizing that information does not circulate solely within neurons, but takes shape in tissues, in fluids, and in slow, profound interactions with the environment. Cognition thus becomes an embodied, distributed, plural process.

Knowing without knowing: the silent intelligence of life.

There is a deeply ingrained prejudice in Western culture: the idea that knowing means thinking, reflecting, and mentally representing the world. And that, therefore, cognition begins only where there is a brain. We are trying to undermine this belief at its root, by showing how life knows long before it thinks—and often without ever doing so. Knowing without thinking is not a poetic provocation, but a biological observation.

As early as the late nineteenth century, on the shores of the Baltic Sea, the Estonian biologist Jakob von Uexküll observed how a tick could remain motionless for years, until three specific signals—the smell of butyric acid, a change in temperature, and the texture of the skin—prompted it to drop onto a passing mammal. No representation of the world, no "reasoning." And yet, an impeccable understanding of what matters. Uexküll called this selective world the Umwelt: not the objective environment, but the one that is meaningful to the organism. To know, in this sense, is to sense what is relevant.

Sensitivity is the first step in this elementary cognition. But it should not be confused with a passive reception of stimuli. A living organism does not register the world like an inert sensor: it intercepts it. In the 1950s, in the laboratories at *Cold Spring Harbor*, some microbiologists noticed that bacteria such as *Escherichia coli* do not move randomly, but modify their behavior based on chemical gradients, anticipating more favorable zones. They do not "think" about nourishment— —

but recognize it even before reaching it. It is a form of temporal knowledge, oriented toward the possible.

This is where the response comes into play. Every living organism not only senses but responds appropriately—that is, in a way consistent with its own survival and identity. The physiologist Claude Bernard, in his Parisian laboratory in the mid-19th century, was struck by the constancy of the internal environment of higher organisms. But what seems even more surprising today is that even organisms lacking a nervous system exhibit integrated responses. Plants, for example. In 1880, Charles Darwin, in his study at Down House, published *"The Power of Movement in* Plants" with his son Francis, describing how roots direct their growth in response to light and tactile stimuli. Darwin spoke, cautiously, of a sort of "sensitive tip" that coordinates the plant's behavior. This insight remained on the margins for a long time, because it challenged the notion that cognition was confined to "higher" animals.

The third dimension—perhaps the most unsettling—is anticipation. To live means to act not only in response to what is happening, but in anticipation of what might happen. In the 1960s, the Italian physiologist Luigi Fantappiè spoke of "biological finality" in dynamic rather than metaphysical terms: living systems organize themselves along oriented trajectories. More recently, studies on circadian rhythms have shown that single-celled organisms anticipate regular environmental variations, such as the light-dark cycle, even when the stimulus is temporarily removed. They do not react to the present; they prepare for the future.

In this light, cognition no longer appears as a late evolutionary luxury, but as a fundamental property of life as an information system. Sensitivity, response, and anticipation are not additional functions: they are the very way in which an organism exists. As the philosopher and physician Georges Canguilhem wrote in 1952, *"to live is to prefer and to exclude."* Every vital act is already an act of cognition, because it discriminates, evaluates, and orients.

Field biology offers a decisive interpretive key here. If the organism is a stabilized information structure, then knowing without thinking means maintaining dynamic coherence with one's own field of relationships. There is no need for a command center, nor for an internal representation of the world. All that is needed is a diffuse sensitivity, capable of transforming information into form, and form into action. Perhaps the real scandal is not that organisms know without thinking. It is that we have long thought we knew without living.

Geese, spinarelli, and instant decisions: cognition without thought.

There was a specific moment in twentieth-century Europe when the idea that thinking is necessary for knowing began to seriously crack. Not in experimental psychology labs, but in meadows, along canals, and in nature reserves. There, where Konrad Lorenz and Nikolaas Tinbergen observed animals doing what they know how to do—without hesitation, without deliberation, and often without ever having "learned" it.

Lorenz recounted this with an ironic smile, recalling his gray geese in Altenberg, near Vienna, in the 1930s. All it took was for him to be the first moving object seen by the chicks that had just hatched for them to follow him everywhere, just as they would their mother. Imprinting is not a cognitive error: it is a form of lightning-fast, irreversible, and perfectly adaptive knowledge. In a world where indecision costs lives, there is no time to think. The organism knows whom to follow because it is wired to recognize certain signals within a precise time frame. Cognition, in this case, is a window that opens and closes.

A few years later, Tinbergen shifted his attention to a seemingly unremarkable fish, the spinarello. In the canals of the Netherlands, he demonstrated that during the breeding season, the male attacks any object bearing a red spot—even a crude cardboard cutout. He does not attack "a rival"; he responds to a key stimulus. The behavior is complex, coordinated, and effective. But it is not mediated by mental representations. It is the response of a sensory system to a meaningful configuration of the perceptual field.

Here, classical ethology takes a decisive step: it shows that the form of behavior precedes intention. Animals do not decide what to do; they recognize when it is time to do it. So-called fixed action patterns—highly organized motor sequences—are activated like prewritten phrases, but only when the environment "utters" the right word. To know, in this sense, is not to construct a model of the world, but to be attuned to certain frequencies within it.

An anecdote often cited in ethology seminars at Oxford in the 1950s tells of Tinbergen, who, during a demonstration, showed a group of students how a female herring gull preferred to peck at a stick with three red stripes rather than the realistic beak of an adult. The exaggerated signal worked better than the original. Not because the animal was "wrong," but because its cognition was tuned to relationships, not to objects. Biological knowledge is always economical: it selects what matters, not what is true in an abstract sense.

This perspective integrates naturally with field biology. If the organism is a stabilized information structure, then behavior is not the execution of an internal program, but the resonance between an environmental configuration and an internal disposition. Sensitivity and response are two sides of the same process. In ethology, anticipation does not appear as conscious prediction, but as structural readiness: the organism is already prepared for what is biologically probable.

Lorenz insisted on a point that is surprisingly relevant today: these behaviors are not rigid automatisms. They can be modulated, diverted, and sometimes inhibited. But their intelligence does not lie in infinite flexibility; it lies in situated precision. An embodied knowledge, inscribed in the organism's form and its evolutionary history. In other words: a cognition that does not think, but is by no means blind.

Viewed from this perspective, classical ethology appears as a frontier discipline that anticipated many contemporary insights. It has shown that knowledge does not coincide with reflective consciousness, that meaning emerges from interaction, and that life, before explaining itself, orients itself. It is a sober yet radical lesson: understanding life means observing how it acts when it has no time to explain why it does so.

Measuring to Live: How Organisms Give Meaning to the World.

If there is one misconception that field biology invites us to overcome, it is the idea that the organism is a mere receptor of stimuli, a sort of control unit that gathers data from the outside world and then reacts. In reality, every living organism measures its own environment. And measuring, in this context, does not mean quantifying according to abstract parameters, but rather establishing meaningful differences and constructing a meaningful world through relationships.

In the 1930s, the Hungarian biophysicist László von Bertalanffy, observing open biological systems, noted that what distinguishes living beings is not the matter of which they are made, but the way in which they maintain stable relationships even amid the continuous flow of energy and information. The organism does not merely endure the environment: it samples it, filters it, and transforms it into a geography of values. Every measurement is already an interpretation.

A seemingly simple example comes from studies on thermoregulation conducted in the 1960s in laboratories in Stockholm. Researchers noted that many mammals do not respond to “objective” temperature, but to rapid, localized variations. In this sense, the skin is not a neutral sensor, but a cognitive surface that weighs changes, distinguishing between what is a threat and what is background noise. The organism measures gradients, not isolated quantities.

This ability emerges with particular clarity in the work of James J. Gibson, a psychologist of perception, who in the 1970s spoke of *"affordance"*: the opportunities for action offered by the environment to a specific organism. A cliff is not "high" in it ; it is either climbable or lethal depending on the body that encounters it. The measure lies neither in the world nor in the organism, but in the relationship between the two. It is here that biological cognition reveals itself as a situated, embodied event, irreducible to an abstract calculation.

An anecdote recounted by an Italian ethologist during an observation campaign in the Gran Paradiso National Park illustrates this point well. While studying chamois, he noticed that the animals altered their usual routes not based on the direct presence of humans, but on micro-variations in the silence and the scent carried by the wind. The environment was "read" as a dynamic map of probabilities. The chamois did not perceive humans, but rather measured an alteration in the relational field in which it lived.

From this perspective, the organism appears as a distributed measuring device, without a single center. Cells, tissues, and organs all participate in this continuous evaluative activity. In the 1990s, Walter Freeman's studies on olfactory dynamics showed that the brain does not recognize odors as fixed imprints, but as configurations that emerge from the interaction between stimulus and internal state. Once again, measurement is not a snapshot, but a process.

Field biology offers a powerful conceptual framework here: the organism stabilizes information not because it stores it, but because it maintains an operational coherence with what it measures. To live is to remain attuned to certain differences while ignoring countless others. Every vital act is an implicit choice, a stance taken within the flow of reality.

As the philosopher Maurice Merleau-Ponty wrote in 1945, *"the body is our general means of having a world."* Not an object among objects, but a measuring threshold, a living interface. In this sense, the active relationship with the environment is not a peripheral chapter of biology: it is its epistemological core. Life does not know the world because it represents it, but because it measures it as it passes through it.

Danger is sensed before it is calculated.

There is a form of intelligence that precedes every calculation, every conscious prediction, every probability chart. It is the intelligence with which organisms assess risk. They do not estimate it, nor do they measure it in percentages: they sense it. Field biology invites us to view this process not as a primitive version of human reasoning, but as an autonomous and refined mode of embodied knowledge.

In this framework, risk is not an external variable to be assessed, but a perturbation of the field of meaning in which the organism lives. A sudden noise, a chemical variation imperceptible to us, or a disruption in the usual rhythm can be enough to trigger a drastic response.

A fascinating example comes from studies conducted in the 1970s by biochemist Howard Berg on bacterial chemotaxis. In his laboratory in Boulder, Berg observed that *Escherichia coli* does not "know" where a food source is located, nor does it calculate the best trajectory. It assesses, moment by moment, whether the situation is improving or worsening. If it worsens, it changes direction. It is a basic yet effective risk assessment: not based on a map of the future, but on a differential sensitivity to the present.

Something similar, but on an infinitely more complex scale, occurs in large herbivores. After wolves were reintroduced to Yellowstone National Park in 1995, ethologists noticed that elk were altering their grazing behaviors even in the absence of visible predators. They avoided certain valleys, changed their schedules, and shortened their stops. The risk was not a direct encounter with a wolf, but the transformation of the entire ecological landscape. The organism gauges danger through the shape of the landscape, not by counting attacks.

In these cases, speaking of a "decision" is misleading. There is no subject weighing alternatives. There is a living system that regulates its own degree of exposure while maintaining internal coherence. The biological measure of risk is always preventive, rarely reactive. Before knowing why something is dangerous, the organism knows that it is.

An anecdote gathered during ethological research on the coast of Brittany, recounted by a French marine biologist, concerns octopuses. Some individuals avoided seemingly ideal burrows after recent storms—not because the burrows had been destroyed, but because the flow of the currents had changed. The risk, for the octopus, was not the den itself, but the loss of control over its own sphere of action. Once again, no estimates, no explicit predictions: just a subtle assessment of the relationship.

This approach to risk assessment is also far removed from the modern, calculation-obsessed human model. In the 1980s, sociologist Ulrich Beck spoke of a "risk society" precisely to describe a culture that attempts to master uncertainty through numbers and simulations. Biological life, on the other hand, seems to follow the opposite logic: reducing exposure while maintaining the quality of the relationship with the environment, not the accuracy of the prediction.

In this light, the biological measurement of risk appears as a fundamental cognitive function, inscribed within the organism itself. It is not necessary to know how likely an event is; it is enough to know whether this situation is still habitable. Survival does not require statistics, but a constant sensitivity to the thresholds beyond which coherence breaks down.

Plants That Decide: How Life Measures Risk Without a Brain.

There is an image that often comes to mind when we speak of cognition: an animal that observes, evaluates, and decides. And then there is the stillness of plants, which for centuries has been mistaken for passivity. Yet, if we look closely, plants are s and masters of a radically different form of knowledge: they do not think, they do not calculate, but they measure. And by measuring, they decide.

As early as 1880, Charles Darwin observed how the root tip behaved "like the brain of a lower animal." It was a provocation, to be sure, but not a mere quip. Darwin had intuited that life, even without a nervous system, possesses a fundamental capacity: to discriminate what is relevant.

For a plant, water is not just a resource: it is a trade-off. Opening the stomata to absorb carbon dioxide means losing water; closing them protects against drought but slows down photosynthesis. It is a constant dilemma, one that renews itself every day, often every hour. In the 1990s, in laboratories at the University of Utah and later at Davis, plant physiologists such as Melvin Tyree and John Sperry demonstrated that many plants operate within what we now call a hydraulic safety margin. In simple terms: they know how far they can "push the limits" before air bubbles—cavitation—form in their vascular tissues, interrupting water flow and potentially leading to tissue death.

This is not a reflex response. Plants anticipate. They adjust stomatal conductance based not only on current moisture levels but also on the system's trend: if drought conditions are worsening, they react before the damage becomes irreversible. In some Mediterranean species, such as the holm oak, this caution is almost proverbial; in others, such as certain lowland poplars, the strategy is riskier, almost "speculative." Two cognitive styles, we might say, embedded in their physiology.

How does a plant "know" that it is taking a risk? There is no command center. The decision-making is distributed. Variations in hydraulic pressure, ionic gradients, and hormones such as abscisic acid travel through the tissues, creating a coherent picture of the situation. Every leaf, every root contributes to the overall assessment.

A researcher explained during a conference in Montpellier in 2008 that working with plants means abandoning the idea of a single decision-maker: *"It's like talking to a crowd that nevertheless responds in unison.*" This is where the notion of an organism as a stabilized information system makes sense: unity is not a point, but a coherence.

Even light, which we instinctively associate with plant life, can become a danger. Excessive radiation damages photosystems, causing oxidative stress. Plants know this—and they react. In the 1950s, in the laboratories of the U.S. Department of Agriculture, Sterling Hendricks and Harry Borthwick discovered phytochromes: pigments sensitive to the quality of light. Thanks to these sensors, a plant can distinguish whether it is in full sun or in the shade of other plants. This is no trivial difference: the "green" shade cast by other plants signals imminent competition.

This gives rise to what is known as *"shade avoidance syndrome*": the stem elongates, the leaves change their angle, and resource allocation is reoriented. It is an assessment of future risk: staying short means being deprived of light; growing quickly exposes the plant to mechanical instability and water stress. Once again, there is no calculation, but rather a dynamic assessment of vital probabilities.

The most surprising thing is that all of this happens without displacement. Or, rather, without what we call movement. Plants gain knowledge by "staying put." Rooted in the soil, they transform time into information. Where an animal can flee, a plant must anticipate.

At the Royal Botanic Gardens, Kew, in London, an elderly curator used to say that plants are "forced philosophers": they cannot act impulsively, so they must be wise. It is an ironic observation, but it captures an essential point. Plant cognition is not a rapid reaction; it is a slow and profound integration of signals.

When we say that every organism knows, we are not attributing a miniature mind to plants. We are expanding the very idea of knowledge. To know, in this sense, means maintaining one's vital coherence in an uncertain world. It means weighing risks and opportunities with the tools that one's form of life makes available.

Plants show us that cognition does not arise from the brain, but from the need to remain intact. They are systems that evaluate, anticipate, and choose—not among ideas, but among biological possibilities. In this silent work of continuous assessment, life reveals its most essential aspect: not a blind mechanism, but a process sensitive to meaning.

Making mistakes to stay alive: the prudent intelligence of life.

There is an idea deeply rooted in our way of thinking—not only in economics, but also in a certain naive understanding of biology—that living means maximizing gain. More growth, more energy, more reproduction. Yet, upon closer inspection, real life seems to follow a different, almost counterintuitive logic: not to maximize, but to avoid making too many mistakes. Or rather, to avoid those mistakes that leave no room for appeal.

In the chapter dedicated to life as a cognitive process, this perspective becomes decisive. If every organism knows, then it also knows its own limits. And error—far from being a mere deviation—becomes a structural element of biological knowledge.

In biology, error is never abstract. It is not a wrong answer on an exam paper, but a choice that can jeopardize an entire life form. This is why organisms do not optimize: they play it safe. As the physiologist Walter Cannon noted as early as the 1930s, while studying homeostasis at Harvard, living systems do not seek maximum performance, but rather stability within limits compatible with survival.

It is a subtle but crucial difference. Maximizing gain exposes one to the risk of catastrophic errors; minimizing the risk, on the other hand, accepts a certain degree of inefficiency. Life, we might say, prefers to be a little boring rather than too daring. A fascinating example comes from the microbial world, where cognition appears in its most essential form. As early as the 1970s, researchers observed in the laboratory that genetically identical populations of *E. coli* bacteria do not all behave the same way when faced with an unstable environment. Some individuals drastically slow down their metabolism, while others continue to grow. This strategy is known as *"bet-hedging*": not staking everything on a single prediction of the future.

Aaron Mitchell, working at Stanford in the early 2000s, showed that this apparent "inefficiency" increases the probability that at least part of the population will survive sudden shocks. Here, the mistake is not avoiding growth when the environment would be favorable, but growing when it will soon no longer be so. Biological wisdom, in this case, consists of not placing too much trust in current conditions.

Even in the most complex systems, such as the human immune system, the same logic holds true. The immune system constantly makes mistakes: it identifies harmless things as dangerous, triggers unnecessary inflammation, and sometimes even triggers autoimmunity. From an engineering standpoint, it's a disaster. From a biological standpoint, it's a sensible choice. In the 1950s, Frank Macfarlane Burnet formulated the theory of clonal selection by observing that the immune system is designed to be hypersensitive. It is better to overreact once than to fail to react when the pathogen is real. Here, too, the epistemology of life is asymmetrical: some errors are tolerable, others are not. Biological cognition does not seek truth, but continuity.

This "wisdom of error" is also clearly evident at the ecological level. In unstable environments—deserts, tundras, intertidal zones—the species that survive are not those that make the most of every favorable window of opportunity, but those that maintain a reserve of possibilities. In the 1980s, field studies conducted in the Negev Desert showed that some plants germinate only a fraction of their seeds even after heavy rains. The rest remain in the soil, waiting. It is an act of distrust toward the present. An Israeli botanist remarked, with a touch of irony, that those plants "do not believe in the promises of the sky." Once again, knowledge does not coincide with optimism, but with a sober assessment of the consequences.

If we return to the idea of an organism as a stabilized informational structure, error takes on a new status. It is not an accident of the cognitive process, but its testing ground. A living system knows because it must act without certainties, and it acts knowing that certain gambles are not permitted. In this sense, avoiding risk is not evolutionary cowardice, but embodied intelligence. Life does not seek the maximum possible, but rather what is reliably sufficient. It is a lesson that challenges many dominant metaphors—from linear progress to algorithmic optimization—and paves the way for a biology of consciousness that does not coincide with performance, but with the responsibility of form.

Life as a measuring tool: when science learns from organisms.

If we look closely at the history of scientific instruments, we discover a curious detail: many of them do not arise from pure mathematical abstraction, but from the—often unconscious—imitation of elementary biological functions. It is as if, even before inventing devices to measure the world, humanity had learned by observing organisms that had already been doing so for millions of years.

In the context of this chapter, where life is described as a cognitive process, this comparison becomes illuminating. The organism, in fact, does not merely react: it measures. And it measures— —always in the interest of its own survival, integrating information, memory, and anticipation. This is exactly what we ask of our scientific instruments. Let's take one of the most iconic instruments of modern science: the thermometer. When Galileo Galilei was experimenting with his thermoscope in Padua in the early 17th century, he was not yet measuring temperature in a quantitative sense, but was recording variations. Hot and cold were perceived differences, not absolute numbers.

It is the same principle that governs a mammal's skin or a reptile's membranes: not knowing "the temperature," but sensing a deviation from a vital condition. The skin is not a neutral sensor: it is a cognitive interface, calibrated to biological thresholds. The modern thermometer does nothing more than make this ancient function stable, repeatable, and communicable.

Even the measurement of time—seemingly the most abstract of all—has deep biological roots. Eighteenth-century marine chronometers, such as the one perfected by John Harrison to solve the problem of longitude, arose from the need for synchronization. But synchronization has always been a capability of living beings.

Every organism possesses internal rhythms: cycles of activity, rest, and renewal. In the 20th century, Jürgen Aschoff's experiments in the Andechs bunker in Germany showed that even in the absence of light and external cues, the human body continues to "keep time." Not with numbers, but with physiological oscillations. The scientific chronometer replicates this ability: maintaining an internal temporal consistency independent of the environment.

Before Evangelista Torricelli invented the barometer in 1643, many animals "knew" when the weather was about to change. Shepherds and sailors had been recounting this for centuries: restless dogs, birds changing course, frogs emerging from the water. Not folklore, but the perception of pressure changes. The barometer does not introduce a new kind of knowledge—

objective knowledge. It transforms a widespread sensitivity into a readable measurement. Once again, the scientific instrument stabilizes a biological function— : detecting significant environmental changes before they become dangerous.

What these examples have in common is a point that is often overlooked: no instrument measures "in absolute terms." Every measurement is a relationship. The rising mercury, the oscillating needle, and the swinging pendulum do nothing more than translate an interaction into a stable signal. Just as an organism does when it perceives, evaluates, and responds. Viewed from this perspective, scientific instruments appear as cognitive prostheses of life itself. Not neutral entities, but extensions of a primary biological capacity: distinguishing what matters from what can be ignored.

Modern science, in its quest for objectivity, has often forgotten this origin. Yet every instrument carries with it an implicit question: what is relevant to a system that seeks to endure? It is the same question that, in a silent form, every organism poses to the world. In this sense, speaking of field biology and of organisms as stabilized informational structures means recognizing that cognition does not begin with the brain, nor with the laboratory. It begins the moment something measures something else in order to remain what it is.

To measure is already to choose: the silent responsibility of the living.

At this point, the question takes on a different tone. It is no longer merely about how organisms measure the world, but what it implies that they do so. Because if life, in all its forms, is a system that measures—selecting signals, filtering out noise, anticipating consequences—then measurement is never an innocent act. It is already a stance.

In the history of science, we are accustomed to thinking of measurement as a neutral act, almost morally transparent. A thermometer does not judge; a sensor does not decide. Yet this d notion falters as soon as we return to the biological origin of all measurement. The organism measures in order to continue existing. Every act of cognition is oriented, situated, and purposeful. In other words: it is already imbued with value. In the 1930s, the physiologist Walter Bradford Cannon, while working at Harvard, observed that the body never reacts randomly to internal and external variations. The concept of homeostasis stems precisely from this: to measure means to maintain conditions compatible with life. But maintaining implies a preference, a norm, a "this way and not otherwise."

This norm is not moral in the human sense of the term, but neither is it ethically neutral. It is a biological proto-ethics: what preserves coherence is favored; what destroys it is avoided. Every organism, in its act of measuring, implicitly establishes what is acceptable and what is not.

This becomes particularly evident when biological measurement is externalized into technology. In 1945, at *the Trinity Site* in the New Mexico desert, measuring instruments recorded the first nuclear explosion with impeccable precision. The Geiger counters worked perfectly. But the ethical issue lay not in the data collected, but in the choice of what to measure and why.

Biology teaches us that no system measures everything. Every measurement is selective. Applying this lesson to science means recognizing that responsibility does not arise after knowledge, but alongside it. As Hans Jonas wrote in **The Imperative of* Responsibility* (1979), the power of modern knowledge compels us toward an anticipatory ethics. But, ultimately, this anticipation is already inscribed in living beings.

There is one aspect that is often overlooked: measuring exposes the one who measures. An organism that perceives becomes vulnerable to the information it receives. Claude Bernard, in his Parisian laboratory on Rue d'Ulm, noted how excessive stimulation can destabilize a physiological system just as much as its absence. Measurement, if not integrated, can lead to disharmony. This observation also applies to contemporary humans , immersed in an overabundance of data. Interdisciplinary biology suggests that responsibility concerns not only the use of information but also the ability to sustain it without losing coherence. A system that measures without integrating fragments.

In this light, ethics no longer appears as a belated addition to knowledge, but as its natural consequence. If every organism is a stabilized informational structure, then responsibility coincides with the care for its own coherence and that of the systems with which it is connected.

This is not yet a matter of justice, rights, or duties in the classical sense. It is something more fundamental: the awareness that knowing means intervening, and that intervening alters the field of which we are a part. When science acknowledges its biological roots, it can no longer evade the question of the meaning and consequences of its actions.

Life, through its very nature, teaches us that all authentic knowledge carries with it a silent responsibility: to remain in relationship without destroying the coherence that makes that relationship possible.

Toward a Biology of Consciousness.

At this point in our journey, after observing life as it measures, selects, anticipates, and maintains its own coherence, a question arises with a certain inevitability: when, and in what way, does all of this become experience? Not yet consciousness in the human, reflective, linguistic sense, but something that resembles it enough to compel us to rethink our categories.

Speaking of the biology of consciousness does not mean searching for a soul hidden within cells nor reducing experience to a molecular algorithm. It means, more soberly but also more radically, asking whether consciousness might not be an emergent property of highly coherent biological fields, just as life emerges when information ceases to be abstract and takes shape. In the 1990s, neurobiologist Antonio Damasio often recounted an episode that took place at *the "Beth Israel Deaconess Medical* Center" in Boston: patients with specific lesions in the emotional areas of the brain were perfectly capable of reasoning but unable to make decisions. Something was missing—not information, but a sense of the value of the alternatives.

This now-classic anecdote becomes particularly interesting when viewed through the lens of field biology. Even before a conscious decision is made, there is a level of global integration that evaluates, guides, and weighs options. It is not thought, but neither is it a simple reaction. It is a field of embodied significance.

In physics, the concept of a threshold is crucial: below a certain intensity, nothing qualitatively new happens; above that threshold, an unexpected property emerges. Some theoretical biologists have begun to wonder whether consciousness functions in the same way. In 2014, during a conference in Tucson dedicated to the foundations of consciousness, there was a lively debate over whether subjective experience was a "switch" or a "gradient." The widespread impression—later recounted informally by several participants—was that consciousness does not appear suddenly, but rather builds up. Where information is sufficiently integrated, stable, and comprehensive, something happens.

In this sense, the biological fields described in the previous chapters can be understood as the breeding ground for experience: dynamic structures that hold the manifold together and make it present to itself. A biology of consciousness does not deny the role of the nervous system, just as the biology of life does not deny chemistry. But it rejects the idea that the brain produces consciousness the way a gland produces a hormone.

As early as the 1940s, the neurologist Kurt Goldstein, working with brain-injured patients in Frankfurt and later in the United States, observed that the organism reacts to injuries as a whole, reorganizing its overall behavior. It is not an isolated function that is lost: the way of being in the world changes. This is a powerful indication that experience cannot be localized like an object but is distributed like a field.

Reaching this point means we have shifted the question about consciousness. No longer "where is it located?", but "when does it emerge?". No longer "what is it made of?", but "what degree of coherence does it require?". It is a shift in perspective that inevitably opens the door to metaphysical questions, yet without abandoning scientific rigor.

The next step will be to question the meaning of this emergence: if consciousness is an event that occurs within the fields of life, then it concerns not only humans, but the very way in which the universe becomes, locally, self-aware. It is here that biology encounters responsibility, and description gives way to meaning.

Consciousness Put in Parentheses: How Biology Lost It and Is Rediscovering It.

For much of the twentieth century, consciousness was an awkward presence in biology. Not openly denied, but carefully avoided. Too subjective, too difficult to measure, too closely tied to that philosophical lexicon that modern science had learned to view with suspicion. If biology was to become a "hard" science, modern science had to rid itself of it. The act of expulsion has a symbolic date: 1913, when John B. Watson published the manifesto of behaviorism. Psychology—and with it, the biology of behavior—would study only what is observable: stimuli and responses. Consciousness? A metaphysical remnant. In American laboratories, from the hallways of Johns Hopkins to the universities of the Midwest, talking about inner experience became almost a breach of scientific etiquette.

The same logic permeated the nascent field of molecular biology. In 1965, Jacques Monod, at the Institut Pasteur in Paris, wrote **Chance and Necessity**. It is a fascinating, rigorous book, yet unequivocal in its message: life is the product of blind mechanisms, and any reference to purpose or inner life is an anthropocentric illusion. Consciousness, in this framework, is not a biological problem, but rather a cultural epiphenomenon. Even Francis Crick, in his office at the Salk Institute in La Jolla, liked to repeat in the 1980s that *"you, your joys and your sorrows, are nothing more than a vast collection of neurons*." A phrase that became famous precisely because it encapsulates the spirit of an era: explaining everything from the ground up, even at the cost of losing something along the way.

Yet something wasn't quite right. Real biological systems—from ecosystems to cells—continued to exhibit properties that defied the sum of their parts. In the 1970s, in Brussels, Ilya

Prigogine studied dissipative structures and demonstrated that order can emerge spontaneously from chaos when a system is far from equilibrium. He wasn't talking about consciousness, but he reopened a door: global organization matters more than individual mechanisms.

Shortly thereafter, in Chile, Humberto Maturana and Francisco Varela introduced the concept of autopoiesis. An organism is not defined by its components, but by the process through which it sustains itself. It is a unit that produces itself and recognizes itself as such. Here, consciousness is not yet explicitly named, but the idea of a system possessing its own perspective begins to take hold.

The return of consciousness to biology did not occur as a sudden twist, but as a slow shift in the cultural climate. By the end of the 20th century, in research centers studying complex systems—from the Santa Fe Institute in New Mexico to European laboratories on nonlinear dynamics—it became increasingly difficult to ignore the problem of experience.

If a system integrates information at a global level, if it maintains dynamic coherence, if it discriminates between what is relevant to its own existence, then the question is no longer whether something akin to consciousness emerges, but when and in what form. Consciousness ceases to be a mysterious object and becomes an emergent property of highly organized biological systems. This historical overview is not intended to nostalgically reevaluate discarded concepts, but to show that biology is shifting its focus. No longer "how to eliminate consciousness from scientific discourse," but "how to describe it without betraying its nature." It is a delicate transition, one that requires new languages and new theoretical responsibilities. The biology of consciousness is not yet a fully developed discipline, but a horizon that is opening up.

XI—Toward a New Biology.

This chapter serves as a threshold. It does not add a new technical piece to the mosaic, but invites us to look at the bigger picture. After tracing the course of life through the threads of coherence, biological fields, embodied information, and distributed cognition, it becomes inevitable to ask what all this means for science—and for us. Chapter XI does not simply conclude the book; it spells out its implications.

Field biology, in fact, is not merely an alternative interpretive model. It is an ontological proposal: it suggests that organisms are not objects among objects, but stabilized informational processes, centers of meaning situated in the world. Accepting this perspective means recognizing that life cannot be reduced to either a machine or chance, without thereby taking refuge in mystical explanations. It is an uncomfortable but fruitful position.

Throughout the twentieth century, science has refined extremely powerful tools, often at the cost of simplifying the object of study. This book has shown how, in the case of living beings, that simplification is beginning to become a limitation. This chapter therefore serves to clarify what changes when life is conceived as a coherent and cognitive field: the legitimate questions and accepted models change, but so does the scientist's epistemic stance. It is not a matter of adding "values" to biology from the outside, but rather of recognizing that every description of living beings carries with it cultural and ethical implications. As Hans Jonas writes in *The Imperative of Responsibility*: *"Act in such a way that the consequences of your action are compatible with the continued existence of an authentically human life on Earth."* This statement, which originated in a philosophical context, takes on a surprisingly biological dimension if life is understood as a network of interdependent fields: intervening in one always means intervening in the whole.

One of the goals of this chapter is to show that the biology of fields does not impoverish science but rather expands it. It introduces new models—systemic, non-local, dynamic—and compels us to question concepts that biology had set aside out of caution: form, purpose, meaning. Not as dogmas, but as legitimate problems.

In this sense, the dialogue with metaphysics is not a step backward, but rather a step to the side. Therefore, this chapter does not offer definitive solutions. It points in a direction. It shows that conceiving of life as an event of meaning implies a more reflective science and a more responsible form of knowledge. It is the point where scientific discourse encounters the question of what is worth knowing—and how to use what we know.

When science stops dismantling and begins to listen.

If field biology is a radical proposal, it is so first and foremost for science itself. Having navigated the crisis of the classical paradigm, the discovery of coherence, and the idea of the organism as a stabilized information field, it becomes inevitable to ask: what really changes in the scientist's work? The answer is not technical, but methodological and cultural. The models change—no longer linear, no longer purely causal—and the questions considered legitimate change as well. Field biology suggests that living systems cannot be understood merely by breaking them down, but must be studied in terms of their organizational patterns, their resonances, and their dynamic stabilities. In other words, science is called upon to observe processes, not just mechanisms.

One of the most evident changes concerns the types of models that are accepted. The dominant metaphor of the machine, though extraordinarily effective at the molecular level, reveals its limitations here. Organisms increasingly appear as coherent systems that maintain their identity through continuous flows of matter, energy, and information. This is not a new idea, but the way it is formalized is changing.

Recall that in the 1950s, Ludwig von Bertalanffy, the founder of systems theory, observed that " " *"The whole is more than the sum of its parts."* A phrase often quoted, but rarely taken entirely seriously. Field biology attempts to do just that, proposing models capable of describing how biological order emerges and is maintained without being localized to a single level. It is no coincidence that conceptual and l tools such as nonlinear dynamics, self-organization, and coherence are now firmly established in the biological lexicon, far beyond the realm of physics where they originated.

Alongside the models, the questions are changing. For decades, biology has prioritized the "how" at the expense of the "why," fearing a drift toward finalism. But a biology of fields cannot avoid questions about the form, stability, and directionality of vital processes—not as external causes, but as immanent properties of living systems.

Conrad Waddington wrote that biological development resembles a landscape of possibilities. Today, that insight is relevant once again: the question is no longer merely which genes are involved, but how the organism maintains its trajectory among many compatible possibilities. It is a question that calls into play regulatory fields, not individual molecules.

Another shift concerns the idea of scientific neutrality. If a living organism is an information system that interprets and responds, then even the act of understanding it is never completely external. Field biology makes explicit what philosophers of science such as Evelyn Fox Keller have long emphasized: models are not mirrors of nature, but situated tools, imbued with metaphors and implicit choices.

Karl Popper noted that *"Science is not a system of certain propositions, but a human endeavor in constant revision."* In this context, the value of a theory lies not only in its predictive power but also in its openness: in its ability to integrate different levels without forcibly reducing them. Field biology calls on science to become more reflective, not less rigorous.

Finally, the relationship between what is measurable and what is meaningful is changing. Not everything that matters in a living system is immediately quantifiable, and not everything that is quantifiable exhausts the phenomenon. This does not mean abandoning measurement, but rather recognizing its limits. As Niels Bohr cautioned, reflecting on complex systems: *"We are not spectators, but actors in the great drama of existence."*

Applied to biology, this statement invites us to develop tools capable of capturing patterns, relationships, and stability, without attempting to reduce them to a single scale. It is a subtle yet profound shift: science does not give up on explaining, but accepts that explaining life also means interpreting it.

What changes for humanity when life is a field?

If field biology redefines what life is, it inevitably redefines who we are as well. Not as a side effect, but as a direct consequence. Because human beings do not observe life from the outside: they are a complex modulation of it, a particular stabilization of embodied information. When the organism is no longer conceived as a machine composed of parts, but as a coherent field that persists over time, identity, health, and our relationship with nature cease to be separate concepts.

From this perspective, identity no longer corresponds to a fixed core—the genome, the brain, the personality—but to a process of dynamic continuity. We are recognizable not because we are unchanging, but because we are consistent. The philosopher Maurice Merleau-Ponty emphasized an idea that was marginal at the time: the body is not an object we possess, but the very way in which we are in the world. Today, that insight finds a precise biological resonance. Conceiving of the human being as a field means recognizing that identity emerges from the constant interplay between levels: molecular, physiological, perceptual, and symbolic.

Health, too, is taking on a new meaning. It is no longer simply the absence of organ dysfunction or the return to normal parameters, but rather the quality of the organism's overall coherence. A healthy body is not one that never deviates, but one that knows how to reorganize itself without losing its unity.

In the 1970s, epidemiologist Aaron Antonovsky—working between Israel and the United States—proposed the concept of "*sense of* coherence" to explain why some people remain healthy under adverse conditions. His insight was that health depended on the ability to give meaning and continuity to experience, not just on isolated biological factors. Today, that idea can be reinterpreted from a biological perspective: health as informational stability, as maintaining one's inner balance.

This also shifts the focus of responsibility: not in the blame-inducing sense of "it's all your fault," but as active participation. Care is no longer just about repair, but about supporting coherence—environmental, relational, and temporal.

Finally, it changes the way humans position themselves within nature. If organisms are fields that interact, then the ecosystem, too, is not merely a container, but a field of fields. Human beings are not above or outside, but within, in constant resonance. When Rachel Carson published **Silent* Spring* in 1962, she chronicled not only an ecological disaster but also a loss of invisible relationships: insects, birds, soils, seasonal cycles. In the book, the author denounces the indiscriminate use of chemical pesticides, showing how they damage ecosystems, kill wildlife, and endanger human health. At the time, she was accused of being overly emotional. Today, she is seen as one of the first voices to intuit that human health and the health of ecosystems are inseparable because they share fields of information and regulation.

James Lovelock, too, while working as a scientific consultant for NASA in the 1960s, arrived at the Gaia hypothesis by observing chemical anomalies in Earth's atmosphere. The Earth, he suggested, behaves like a self-regulating system—not a deity, but a coherent organization. Within this framework, humans are not rulers, but powerful modulators—and therefore responsible.

Ultimately, what changes for humanity is not a single belief, but the underlying tone. The idea of being separate entities gives way to that of being sensitive nodes in a living web. This does not diminish individuality; it deepens it. It does not dissolve freedom; it anchors it to real conditions. It does not idealize nature, but recognizes its shared complexity. As Gregory

Bateson wrote, reflecting on living systems: *"The severing of connections is always a form of madness.*" Field biology suggests that human health—biological, psychological, and cultural—depends on the quality of the connections we are able to maintain.

Science and Metaphysics Reconciled.

For a long time, in Western culture, science and metaphysics have viewed each other with suspicion. The former accused the latter of vagueness; the latter reproached the former for being blind to meaning. Field biology, almost unwittingly, is now forcing both to reconsider their positions. Not to merge into a confused syncretism, but to recognize that living beings pose questions that neither, on its own, can any longer address.

The starting point is surprisingly concrete. When life is described as a stabilized information structure—as a coherent field that organizes matter and energy over time—purely mechanistic language begins to show cracks. Not because it is false, but because it is incomplete.

Since then, the science of complexity has made enormous strides, but the philosophical crux remains. Ilya Prigogine liked to repeat that *"the future is not given."* He said this while studying systems far from equilibrium, not while writing existential essays. Yet that phrase has an obvious metaphysical significance: if nature produces real novelty, if time is creative, then the world is not merely an unfolding of predetermined causes. Field biology takes up this legacy and brings it into the very heart of life.

This is where reconciliation takes place. Metaphysics does not re-enter through the door of ineffable mystery, but through that of structure. It no longer speaks of "vital forces" in a nebulous sense, but rather of principles of organization, of unity that precedes the parts, of meaning as an intrinsic dimension of living systems. It is a sober metaphysics, we might say, that renounces " " ultimate explanations but rejects the idea that meaning is a subjective illusion.

An anecdote often recounted in French academic circles tells of a discussion that took place in Paris in the late 1980s between a molecular biologist and a philosopher of science following a lecture at the Musée de l'Homme. The former argued that *"in the end, everything boils down to molecules*." The latter replied: *"Yes, but not everything is decided there*." It is in this slight but decisive difference that the biology of fields finds its place.

Reconciliation also takes place on the level of the responsibility of knowledge. Hans Jonas, in **The Imperative of* Responsibility* (1979), did not speak of biological fields, but he had lucidly grasped the problem: a highly powerful science, devoid of reflection on meaning and consequences, becomes dangerously short-sighted. If life is a network of informational fields, every technical intervention is never merely local but resonates on levels that we often cannot immediately measure. Metaphysics, here, does not dictate laws to science: it reminds science of the breadth of its impact.

It is important to clarify what this reconciliation is not. It is not a return to mysticism, nor an improper spiritualization of biology. But neither is it a refined form of reductionism that changes the vocabulary while leaving the underlying framework intact. Speaking of fields means accepting that there are levels of reality describable only in relational and formal terms, not fully translatable into linear causal chains.

In this sense, field biology plays a cultural role similar to that of general relativity in the early twentieth century. When Einstein presented his equations in Berlin in 1915, he proposed not only a new physics but a new way of thinking about space and time. Similarly, a biology that recognizes stabilized information fields compels us to rethink categories such as cause, form, emergence, and purpose—words that science had expelled but which now return, transformed and kept in check.

The result is not a definitive synthesis, but a fruitful coexistence. Science continues to do what it does best: observe, measure, and model. Metaphysics renounces total explanations, but returns to questioning assumptions, limits, and overall meaning. Life, meanwhile, remains there: concrete, measurable, yet irreducible to a mere sum of data.

Beyond explanation, before mystery: life as an event that speaks.

At this point, after having explored the crises of the paradigm, coherence, fields, health, and biological knowledge, one question remains that can no longer be avoided. It is not a technical question, nor is it a methodological problem. It is a question of status: what kind of reality is life, if to understand it we must invoke not only reactions, selections, and information, but also unity, form, orientation, and meaning?

Field biology does not provide a definitive answer, but it makes it clear that the living is not neutral with respect to meaning. Not in the sense that it "means something" like a symbolic text, but in the more sober and profound sense: life unfolds as a structured, coherent, oriented event. And where there is orientation—even minimal, even impersonal—a question of meaning inevitably arises.

A physicist like John Archibald Wheeler, in the final years of his life, liked to summarize his vision with a curious formula: *"it from bit."* Physical reality, he suggested, emerges from information. It was a bold hypothesis, but one that revealed a cultural climate: when science reaches the heart of processes, it encounters something that is not merely matter in motion, but configuration, a choice among possibilities, and the stabilization of forms. Field biology brings this insight into the realm of the living, showing that organisms are not merely products of the world, but events that organize the world locally.

Here, science and metaphysics cease to vie for territory and finally begin to acknowledge one another. Science accepts that the description of the living involves levels of reality that cannot be exhausted by linear causality. Metaphysics, in turn, refrains from speaking in place of science and limits itself to questioning what science presupposes without being able to measure it directly: unity, identity, direction, value.

Conceiving of life as an event of meaning does not mean attributing to it an external purpose or a predetermined cosmic plan. It means recognizing that, in the known universe, life is the place where information becomes organized experience, where matter ceases to be indifferent to itself. As the philosopher Hannah Arendt wrote, reflecting on birth, every new beginning *"introduces into the world something that was not there before.*" The biology of fields shows that this applies not only to human action but to living beings as such.

This chapter thus concludes at a threshold. It does not offer a metaphysics of life, but it shows why a reflection on meaning is not an arbitrary addition, but rather a natural consequence of what contemporary biology is discovering. If life is a coherent field that persists over time, if it knows, measures, selects, and responds, then it is not merely a fact of the universe, but an event that transforms it from within.

The book's conclusion can now openly address this implication: not what life means, but why, at the very heart of nature, meaning becomes a real possibility.

Conclusion. Life as an Event of Meaning in the Universe.

Having reached this conclusion, the path taken by field biology allows us to formulate a statement that, just a few decades ago, would have seemed out of place in a scientific context: life is an event of meaning in the universe. Not a poetic addition, not a philosophical consolation, but a rigorous consequence of what emerges when the living is observed as a stabilized informational structure, as a coherent field that organizes matter, energy, and information over time.

For a long time, the dominant idea was that life was an improbable deviation in a fundamentally indifferent cosmos. Jacques Monod, in his famous *Chance and Necessity*, spoke of humanity's fundamental "solitude" in a blind universe. That vision had enormous cultural influence, but today it appears to be a product of its historical context. Not because it has been "disproven," but because our framework of understanding has changed. By studying complex systems, self-organization, and nonlinear dynamics, it becomes increasingly difficult to maintain that life is merely a statistical accident.

This book suggests another perspective: life as a natural mode through which the universe, under certain conditions, produces lasting coherence and embodied information. It is not a supernatural event, nor is it a mere mechanical consequence. It is a phase transition. Just as water, at a certain temperature, ceases to be liquid and becomes ice with new properties, so too does matter, by organizing itself into biological fields, generate something that did not exist before: a center of relationship, memory, and response.

A little-known anecdote concerns a lecture given by Freeman Dyson at Princeton in the early 2000s. At the end of the talk, a student asked him if he considered life "inevitable" in the universe. Dyson smiled and replied, *"I don't know if it's inevitable, but it certainly isn't surprising.*" In that statement lies a subtle yet decisive shift: life is not a scandalous exception, but rather a coherent outcome of a cosmos rich in organizational possibilities.

Conceiving of life as an event of meaning does not imply attributing a predetermined purpose to it. Meaning, here, is not an end imposed from the outside, but an emergent property. Every living organism selects, interprets, and distinguishes what matters from what does not. Even a bacterium, in its microscopic environment, lives in a meaningful world, made up of gradients, thresholds, and signals. Jakob von Uexküll had already intuited this in the 1920s when he spoke of the Umwelt: every living being inhabits a world of meaning, not a neutral universe.

If this holds true for every organism, then the emergence of life introduces a new quality into the universe: the possibility that something matters. Not to anyone in particular, not to an external observer, but to the living system itself. *"The Biology of* Fields" makes this transition visible, showing how biological coherence is always also functional coherence, oriented toward persistence, relationship, and continuity.

An anecdote recounted by a primary care physician during the difficult years of the pandemic illuminates this point from a concrete perspective. Faced with patients with similar clinical parameters, some responded to treatment, while others did not. "*At a certain point,*" he said, "*I realized that I wasn't just treating bodies, but stories trying to remain whole.*" This is not a scientific statement in the strict sense, but it is an intuition that resonates deeply with the idea of a field: life as a unit of meaning that strives to sustain itself.

In this light, the universe no longer appears as an indifferent backdrop against which life happens by accident. Rather, it appears as a context capable, at least locally, of producing meaning through life. The biology of fields does not claim that the universe "wants" life, but shows that the universe can become meaningful thanks to life. It is a subtle but crucial difference.

Concluding by speaking of meaning does not mean stepping outside of science, but rather recognizing its fruitful limit. Science describes how life happens; meaning emerges from the very fact that it happens this way, and not otherwise. In this space—not empty but open—lies human responsibility: to understand that, by studying life, we are observing one of the ways in which the universe becomes interpretable, even narratable.

Life, then, is neither a miracle that violates the laws of nature nor an insignificant accident. It is one of the events in which nature, following its own laws, overcomes its own indifference and ushers in a world where something matters.

1 Bibliography.

Below is an extensive specialized bibliography consistent with the book's conceptual framework, focusing on fields such as biology, information, form, coherence, complexity, and beyond reductionism, and including classics, scientific texts, theoretical works, and interdisciplinary studies. Titles are listed in their original language, when available.

Abrams David	The Spell of the Sensuous – Pantheon Books
Bateson Gregory	Mind and Nature: A Necessary Unity – Dutton
Bateson Gregory	Steps to an Ecology of Mind – University of Chicago Press
Behe Michael J.	Darwin's Black Box – Free Press
Bohm David	Wholeness and the Implicate Order – Routledge
Bohm David	Causality and Chance in Modern Physics – Routledge
Capra Fritjof	The Web of Life – Anchor Books
Capra Fritjof, Luisi Pier Luigi	The Systems View of Life – Cambridge University Press
Davies Paul	The Cosmic Blueprint – Simon & Schuster
Davies Paul	The Fifth Miracle – Simon & Schuster
Deacon Terrence W.	Incomplete Nature: How Mind Emerged from Matter – W.W. Norton
Del Giudice Emilio	Role of Coherence in Biological Systems – World Scientific
Elsasser Walter	The Physical Foundation of Biology – Pergamon Press
Elsasser Walter	Atom and Organism – Princeton University Press
Goodwin Brian	How the Leopard Changed Its Spots – Princeton University Press
Goodwin Brian	Nature's Due – Floris Books
Gould Stephen Jay	The Structure of Evolutionary Theory – Harvard University Press
Hameroff Stuart, Penrose Roger	Consciousness and the Universe – Cosmology Science Publishers

Ho Mae-Wan	The Rainbow and the Worm – World Scientific
Hoffmeyer Jesper	Signs of Meaning in the Universe – Indiana University Press
Hoffmeyer Jesper	Biosemiotics – University of Scranton Press
Jantsch Erich	The Self-Organizing Universe – Pergamon Press
Kauffman Stuart	The Origins of Order – Oxford University Press
Kauffman Stuart	At Home in the Universe – Oxford University Press
Keppler Joachim	The Energetic Basis of Life – Springer
Lovelock James	Gaia: A New Look at Life on Earth – Oxford University Press
Margulis Lynn	Symbiotic Planet – Basic Books
Margulis Lynn, Sagan Dorion	What Is Life? – University of California Press
Maturana Humberto, Varela Francisco	Autopoiesis and Cognition – D. Reidel
Maturana Humberto, Varela Francisco	The Tree of Knowledge – Shambhala
Montagnier Luc	DNA Waves and Water – AIP Conference Proceedings
Morin Edgar	La Méthode – Seuil
Pattee Howard	Hierarchy Theory – George Braziller
Penrose Roger	The Emperor's New Mind – Oxford University Press
Prigogine Ilya	From Being to Becoming – W.H. Freeman
Prigogine Ilya, Stengers Isabelle	Order Out of Chaos – Bantam Books
Rosen Robert	Life Itself – Columbia University Press
Rosen Robert	Essays on Life Itself – Columbia University Press
Sheldrake Rupert	A New Science of Life – Blond & Briggs
Sheldrake Rupert	The Presence of the Past – Times Books
Stengers Isabelle	Thinking with Whitehead – Harvard University Press
Uexküll Jakob von	A Foray into the Worlds of Animals and Humans – University of Minnesota Press
Varela Francisco	Ethical Know-How – Stanford University Press
Waddington C.H.	The Strategy of the Genes – Allen & Unwin
Whitehead Alfred North	Process and Reality – Free Press
Wiener Norbert	Cybernetics – MIT Press
Wolfram Stephen	A New Kind of Science – Wolfram Media

Studies and Research

Below is a specialized list of recent and recognized scientific studies and projects consistent with the book's theme.

I have prioritized works on biological coherence, fields, information, non-locality, complex systems, biophysics, and embodied cognition, while avoiding purely metaphorical titles.

The titles are in their original language; the list is arranged alphabetically by the author's last name.

Albert Rémi	Institut de Biologie de l'École Normale Supérieure	*Network Inference, Analysis, and Modeling in Systems Biology* – Nature Reviews Genetics
Bialek William	Princeton University	*Biophysics: Searching for Principles* – Princeton University Press
Cifra Michal	Czech Academy of Sciences	*Electromagnetic Fields Generated by Living Cells* – BioSystems
Del Giudice Emilio	INFN – Istituto Nazionale di Fisica Nucleare	*Coherent Quantum Electrodynamics in Living Matter* – Journal of Physics: Conference Series
De Loof Arnold	University of Leuven	*The Cell's Self-Generated Electromagnetic Field* – Communicative & Integrative Biology
Ellis George F.R.	University of Cape Town	*Top-Down Causation and the Emergence of Information in Complex Systems* – Interface Focus
Fields Chris	Independent Researcher	*How Do Living Systems Create Meaning from Information?* – Biosemiotics
Fröhlich Herbert	University of Liverpool	*Long-Range Coherence and Energy Storage in Biological*

		Systems – International Journal of Quantum Chemistry
Giudice Del Emilio, Preparata Giuliano	INFN	*Biological Ordering and Coherent Domains in Water* – Journal of Biological Physics
Haken Hermann	University of Stuttgart	*Synergetics: Introduction and Advanced Topics* – Springer
Ho Mae-Wan	Institute of Science in Society	*Quantum Coherence and the Living Organism* – BioSystems
Keppler Joachim	University of Tübingen	*The Energetic Basis of Information and Consciousness* – Entropy
Kauffman Stuart	Institute for Systems Biology	*Autocatalytic Sets and the Origin of Life* – Journal of Theoretical Biology
Levin Michael	Tufts University – Allen Discovery Center	*The Computational Boundary of a "Self"* – Frontiers in Psychology
Levin Michael	Tufts University – Allen Discovery Center	*Bioelectric Signaling: Reprogrammable Circuits Underlying Embryogenesis* – Cell
Longo Giuseppe	CNRS – École Normale Supérieure	*Extended Criticality and the Physics of Life* – Journal of Biological Systems
Montagnier Luc	Fondation Luc Montagnier	*DNA Waves and Water* – Interdisciplinary Sciences: Computational Life Sciences
Mossio Matteo	CNRS – Sorbonne Université	*Organizational Closure in Biological Organisms* – Journal of Theoretical Biology
Noble Denis	University of Oxford	*Biophysics and Systems Biology: Why Reductionism Fails* – Journal of Physiology
Popp Fritz-Albert	International Institute of Biophysics	*Biophoton Emission: New Evidence for Coherence in Living Systems* – Physics Letters A
Prigogine Ilya	Université Libre de Bruxelles	*Non-Equilibrium Systems and the Emergence of Order* – Physics Today
Rosen Robert	Dalhousie University	*Life Itself: A Comprehensive Inquiry into the Nature, Origin, and Fabrication of Life* – Columbia University Press
Simeonov Plamen	BAS – Bulgarian Academy of Sciences	*Integral Biomathics: A Post-Newtonian View into the Logos of Bios* – Springer

Torday John	UCLA	*Evolutionary Biology, Cell Communication and the Origin of Physiological Coherence* – Progress in Biophysics and Molecular Biology
Ulanowicz Robert	University of Maryland	*Information Theory in Ecology and Biology* – BioSystems
Varela Francisco	CNRS – CREA, École Polytechnique	*The Brainweb: Phase Synchronization and Large-Scale Integration* – Nature Reviews Neuroscience
Walker Sara Imari	Arizona State University	*The Information Theory of Life* – Reports on Progress in Physics
Werner Eric	University of Oxford	*The Origin, Evolution and Development of Biological Complexity* – BioSystems

Back Cover

Are we really just the result of billions of chemical reactions orchestrated by DNA? Or does life conceal a deeper organizing principle, capable of giving form, coherence, and meaning to living matter? Field Biology invites the reader to question one of the most deeply rooted certainties of modern science: the idea that understanding the parts is equivalent to understanding the whole.

Written in an accessible yet rigorous style, this book guides the reader through the crisis of the classical biological paradigm—where biochemical explanations reveal their limitations and Darwinism struggles to account for the problem of form. How can an organism maintain its identity, develop coherently, and adapt creatively to its environment if there is no principle coordinating the parts even before they come into play?

The proposed answer is as bold as it is fascinating: life as a phenomenon of coherence. From the physics of complex and quantum systems to biology, the image of the organism emerges as a non-local system, in which the whole precedes and informs the parts. Within this framework, biological fields take shape: not vague metaphors, but real informational structures capable of stabilizing form, guiding vital processes, and making living matter something more than a random assemblage.

Information that takes on physical form, vital energy to be reimagined in scientific terms, health as a state of global coherence, and disease as a fragmentation of the field: each chapter opens a new perspective on living and knowing. Life thus appears as a diffuse cognitive process, in which every organism—, from the simplest to the most complex—interprets the world and responds to it in a meaningful way.

Field biology does not promise definitive formulas, but offers an integrated and responsible vision, capable of bringing science, philosophy, and tradition into dialogue without confusing them. A book for those who feel that reducing life to a mechanism is no longer enough, and who feel the need to conceive of the organism as an event of meaning in the universe.

Short Synopsis ISBN

This book offers an innovative reinterpretation of the phenomenon of life, aimed at those who recognize the limitations of a biology reduced to mere chemistry and genetics. Written in an accessible yet rigorous style, the book explores the crisis of the classical paradigm, demonstrating why life cannot be explained as a simple sum of reactions or as the blind outcome of natural selection. At the heart of the work lies the concept of coherence: life as a unified process in which the whole precedes and organizes the parts. The organism is thus described as a non-local system, sustained by biological fields understood not as metaphors but as real informational structures, capable of guiding form, stabilizing identity, and integrating matter, energy, and information. This perspective opens up new ways of understanding health and disease—conceived, respectively, as states of coherence or fragmentation of the vital field—and leads to a vision of life as a diffuse cognitive process. A book for readers interested in a biology capable of engaging with philosophy and restoring life's value as a source of meaning in the universe.

The Author

Bruno Del Medico, born in 1946 in Sabaudia (Italy), spent his professional career working in computer science and software development. He subsequently devoted himself to the study and popularization of the most current scientific theories, such as quantum theory in its metaphysical sense. Bruno Del Medico is currently the author of numerous works published by Edizioni PensareDiverso. Bruno Del Medico's work stands out from traditional narratives in the field of technical sciences due to his critical, independent, and unconventional approach. Instead of limiting himself to standard interpretations or the general consensus on purely scientific topics, Bruno Del Medico explores alternative perspectives that often go against the grain, challenging commonly accepted assumptions. A distinctive feature of his works is his ability to weave together science, ethics, and philosophy in an open dialogue, which encourages readers to reflect independently and consider the deeper implications of technological and scientific innovations.

life as an informational process – vital fields,
organisms as coherent systems – nonlocality,
biology and quantum physics – coherence,
embodied information – biological form,
health and disease – loss of coherence,
epistemology of biology – new science,
life and meaning – the metaphysics of nature,

AM – field biology – life as information,
new biology – beyond reductionism,
biological coherence – physics and life,
biological fields – the non-local organism,
health as coherence – systemic disease,
Darwinism and form – a critique of evolutionism,
quantum biology – complex systems,

www.ingramcontent.com/pod-product-compliance
Lightning Source LLC
LaVergne TN
LVHW052028200926
848481LV00005B/1611